Springer Praxis Books

# Popular Science

The Springer Praxis Popular Science series contains fascinating stories from around the world and across many different disciplines. The titles in this series are written with the educated lay reader in mind, approaching nitty-gritty science in an engaging, yet digestible way. Authored by active scholars, researchers, and industry professionals, the books herein offer far-ranging and unique perspectives, exploring realms as distant as Antarctica or as abstract as consciousness itself, as modern as the Information Age or as old our planet Earth. The books are illustrative in their approach and feature essential mathematics only where necessary. They are a perfect read for those with a curious mind who wish to expand their understanding of the vast world of science.

More information about this subseries at http://www.springer.com/series/8158

Paul J. Hazell

# The Story of the Gun

History, Science, and Impact on Society

 Springer

Published in association with
**Praxis Publishing**
Chichester, UK

Paul J. Hazell
School of Engineering and Information Technology
UNSW Canberra
Australian Defence Force Academy
Canberra, ACT, Australia

Springer Praxis Books
ISSN 2626-6113            ISSN 2626-6121   (electronic)
Popular Science
ISBN 978-3-030-73651-4       ISBN 978-3-030-73652-1   (eBook)
https://doi.org/10.1007/978-3-030-73652-1

This Springer imprint is published by the registered company Springer Nature Switzerland AG
The registered company address is: Gewerbestrasse 11, 6330 Cham, Switzerland

*To Harold Roberts (1923–)*

*D-Day veteran and gunner*

*And most importantly, my grandfather*

# Preface

My first encounter with a gun was striking—well it was to me, at least. I was part of an aid convoy organised by a Christian charity on the way to Romania a couple of years after the demise of Nicolae Ceaușescu. Romania was in disarray at that time and the people in the villages were desperately poor. I was young, straight out of University, and idealistic. We crossed the border between Hungary and Romania near Oradea. Due to Romania's poverty, the border guards were poorly paid. I remember feeling a little threatened as the guards brandished AK47 assault rifles in an attempt to garner a small gift (some would call it a bribe) from those who crossed the border. Of course, we obliged. Seeing the assault rifles casually slung around the border guards' necks was chilling to me—particularly as I came from a country where guns were never (or only rarely) seen in public. In fact, I think this was the first time that a saw a real-life assault rifle! I do not think they were ever pointed at us in anger, but their presence sent a very cold message to travellers. Little did I realise that I would spend my next career-move thinking about the implications of bullet impacts. And, I have been doing that for the past 25 years.

It seems every day that guns make the news. This is not surprising given that, by one estimate, there are over 1 billion firearms in circulation, 85% of which are owned by civilians. We have become accustomed to hearing about gun-related robberies, mass shootings, suicides, accidental shootings and so on. Our entertainment is full of them too—from computer games such as 'Call of Duty' (and yes, I used to enjoy the odd first-person shooter from time to time) to the Hollywood blockbuster (of which there are too many to list). The fact is, society is *obsessed* with guns and this is perhaps why I wanted to write this book. So, in this book I wish to outline the science behind gun operation and perhaps dispel some of the myths along the way. We will be reviewing some history too. In particular I want to take you down the road of some of the developments in gun and weapons technology that have occurred over the past 300+ years. And, I will not just be focussing on small guns—but I will be discussing the big guns and fast guns too. Finally, we will be looking at the challenging subject of gun control and whether gun ownership is harmful to families and society as a whole.

This is intended to be a book that is accessible to all audiences. I have avoided detailed mathematical descriptions that would be found in student reference and textbooks. My hope is that irrespective of your technical background, you will be able to get something out of this text.

Canberra, Australia                                    Paul J. Hazell
January 2021

# Acknowledgements

I would like to acknowledge many of the students that I have taught over the years, many of whom who are serving military personnel with first-hand experience of being under fire. I have learnt much through our out-of-class discussions on the science and use of guns. I would also like to acknowledge Dr. Heath Pratt for his useful discussions and review of Chap. 2, my colleague Lt. Col. Mick Cook from the Defence Research Institute for helpful reviews of Chaps. 3 and 4, Dr. Ali Ameri for his useful comments on Chap. 5, Dr. Adrian Garrido Sanchis for his support with Chap. 6, Dr. Joe Backofen Jr. for his thorough review of Chap. 7 as well as Prof. Tom Frame AM for his generous discussions and help with Chap. 9. Finally, I would also like to acknowledge my colleagues at UNSW Canberra as well as the wider Defence community with whom I share a passion for all things to do with 'impact'. We have the best jobs in the world!

# Contents

# Chapter 1
# The Birth of the Gun

Guns have been around for centuries. In fact, there has not been many advances in gun technology in the past 150 years. Sure, we know that the ergonomic features of guns have improved, as have their ability to fire projectiles to very high velocities but their basic function have not changed since the Chinese invented gunpowder. In short, we take an energetic material (i.e., propellant) that we ignite and use to accelerate a projectile along a barrel towards its target. And, yes, of course velocities and payloads have improved. Today a high-velocity rifle can fire projectiles to velocities of 950 m/s+. That is 2125 mph or around 2.8 times the speed-of-sound. The accelerations that a modern bullet experience are around 100,000 g—awesome by any standard. So where did it all begin?

## 1.1 So How Did We Get Here?

It seems clear that since the very beginning when people had an inkling to hurt others that it would be better to throw a projectile than to be standing close to the enemy. The reason would have been obvious: the further you are away from the person that would cause you harm, the better! We can throw projectiles at quite incredible speeds. In fact, the human body is structured so that it has the capacity to act as quite a powerful war machine, even with the absence of an Iron-Man suit! In 2010, Aroldis Chapman managed to clock a speed of 105.1 mph (169.1 km/h) by throwing a baseball, which is the fastest baseball pitch in history. The game of cricket has produced a number of fast bowlers over the years. The record for the fastest ever bowled ball was 100.2 mph (161.3 km/h) by Shoaib Akhtar (Pakistan) against England in 2003. These velocities give us an idea of what humans can achieve without any mechanical means. And all this from the chemical energy derived from food and drink used in conjunction with well-honed muscles and well-learned technique. It was soon realised that some form of mechanical advantage would improve the speed and therefore range and lethality of the projectile. For example, ancient slings were

© Springer Nature Switzerland AG 2021
P. J. Hazell, *The Story of the Gun*, Springer Praxis Books,
https://doi.org/10.1007/978-3-030-73652-1_1

used (as per the Biblical account of David versus Goliath in 1 Samuel 17) and these would have been capable of launching projectiles to many tens-of-metres-per-second. By all accounts, the projectiles could be catapulted up to a quarter of a mile and a skilled operative could be accurate to over 200 yards (183 m). And then there is the bow and arrow. A strong long-bow man could easily achieve velocities approaching a couple hundred metres-per-second. And to achieve this took some training. English long-bow men, for example, developed extra-large trapezoid muscles so they could launch their arrows with deadly speed. The longbow was probably the first method for substantially increasing mechanical advantage for launching a projectile by a single person. This was achieved as follows: energy, which is the capacity to do work, was stored in the limbs of the bow as the string was drawn rearward. This energy is called elastic strain energy (strain is a measure of stretching). As the string was released, the energy stored in the limbs of the bow was transferred to the arrow. The arrow was accelerated and eventually reached a maximum velocity at the point the string stopped acting on it. From then on, the arrow was decelerated by drag forces as it flew through the air. The power that was delivered by rapidly releasing the elastic energy that was stored in a bow far outweighed the power that could be derived from fast-twitch muscles throwing the missile. Ironically, the understanding of stored energy was not fully realised until Robert Hooke (1635–1703) came along several hundred years later. Hooke's law stated that the deflection of a material is directly proportional to the force that is applied to that material. The law laid the basis for our understanding of stress and strain and stored energy. It also underpinned our understanding of elastic materials and explained why bows could be used time and time again without failure.

For the bow, and of course catapult and crossbow, elastic strain energy was stored by some mechanical action of the operators. This was accomplished by pulling back on a bow string, often made from hemp or other vegetable fibre, which in turn placed the yew bow in tension. Remarkably, in the medieval ages the whole of the male English population was expected to be involved in warfare. In 1181 the 'Assize of Arms' was passed which was a proclamation of King Henry II of England that every man between the ages of 15 and 60 years old were required to equip themselves with a bow and a selection of arrows. Under Edward IV, every Englishman was to have a bow of his own height, made from yew, wych, hazel, or ash, according to his strength. The arrows themselves were to be the length of a man's arm or half the length of a bow. The long bow for English warfare was so important that during the reign of Henry VII the use of any other bow apart from the long bow was forbidden. In the following reign, a fine of 10 lb, an extraordinary sum of money, was to be paid by whoever might be found to possess a crossbow. Arguably this was the beginnings of what we understand today as 'weapon control' (notwithstanding modern firearms were yet to be invented) although the main focus was probably to ensure that prowess with a longbow was maintained. Nowadays the notion of arming an entire population would seem bizarre. In fact, governments tend to take great care in controlling weapons of war. More on that later.

For larger ranged attacks on fortifications, trebuchets, catapults and ballistas or 'engines' became famous! Again, the notion of stored energy is important here.

Arguably the first mention of such engines can be found in the Old Testament of the Bible. It is noted that in 2 Chronicles 26:15 that King Uzziah (ca. 807–740 BC) *"made devices invented for use on the towers and on the corner defences so that soldiers could shoot arrows and hurl large stones from the walls."* One of the more shocking historical accounts of the use of such engines comes from Josephus, describing a night-time catapult bombardment during Vespasian's siege of Jotapata during the Jewish Revolt in 67 A.D.:

> "The force of the spear-throwers and catapults was such that a single projectile ran through a row of men, and the momentum of the stones hurled by 'the engine' carried away battlements and knocked the corners off towers. There is in fact no body of men so strong that it cannot be laid low to the last rank by the impact of these huge stones. The effectiveness of 'the engine' can be gathered from incidents of that night: One of the men standing near Josephus on the rampart got into the line of fire and had his head knocked off by a stone, his skull being flung like a pebble from a sling some 600 yards; and when a pregnant woman was struck in the belly upon leaving her house at daybreak, the unborn child was carried away 100 yards; so tremendous was the power of that stone-thrower. Even more terrifying than the actual engines and their missiles was the rushing sound and the final crash. There was a constant thud of dead bodies as they were thrown one after another from the rampart (Williamson 1976).

Ballistas were a slight improvement on the catapult and used the energy stored in torsionally-loaded springs to accelerate a bolt to its target. In fact, the origin of the English word *gun* is thought to come from the name of a remarkably large ballista that was installed in Windsor castle during the 14th C (Domina Gunilda[1]). Ballistas were probably of Greek origin (ballō = Greek for 'throw').

Trebuchets, on the other hand, relied on flinging a projectile, usually a "gunstone" to great distances by using the mechanical advantage of a heavy weight attached to lever. A modern reconstruction made in England has thrown a compact car (476 kg without its engine) 80 m using a 30 ton counterweight and so it is understandable why they survived well after the invention of gunpowder (Eigenbrod et al. 1995). Trebuchets tended to be preferred over and above catapults due to their ability to throw a projectile to a greater range and to a better accuracy. The projectile did not need to be a hard, penetrating object either. Often carcasses of dead horses, slain comrades and even living prisoners were propelled over castle walls. The trebuchet also heralded mankind's first attempt at biological warfare. In 1346 at the Siege of Caffa, diseased cadavers carrying the Black Death, possibly many thousands, were hurled into the city. The defenders had to handle the cadavers and therefore exposed to the plague (Wheelis 2002). Further, at the siege of Carolstein in 1422, the defenders were, by all accounts, bombarded by two-thousand cartloads of manure! Arguably, this was an unwitting, if not smelly attempt at biological warfare!

Trebuchets worked by converting the potential energy of an elevated counterweight to the kinetic energy[2] of a projectile. This was achieved as the counterweight fell towards the ground. Potential energy is a measure of the energy an object

---

[1]This name was probably derived from the old Norse woman's name Gunnhildr which combines two Norse words referring to battle.

[2]Kinetic energy is the energy a projectile possesses due to it being in motion.

possesses by virtue of its position relative to the centre of the Earth[3]. The heavier or higher the counterweight, the larger its potential energy and therefore the higher the velocity of launch. The higher the velocity of launch, the greater the range. In addition, the closer the counterweight to the pivot point or fulcrum, the heavier it needed to be to achieve a high velocity. Therefore, these machines could be quite massive.

As with these weapons, where stored energy (elastic or potential) was important, it was the stored energy available in gunpowder, and later nitrocellulose-based propellants that led to guns. You see, if it was not for the stored chemical energy present within many chemical substances, then guns would have never arrived on planet Earth. Neither would have locomotion, motor cars, transportation, electricity and so on, for that matter!

## 1.2  Beginnings—A Brief History of Gun Developments

To propel a projectile to high velocity there is a need for a chemical substance that can release energy sufficiently quickly, that in itself will not break the surrounding chamber. It is merely a process of energy conservation. That is, the chemical energy is converted to kinetic energy by virtue of an exothermic reaction. That is, a chemical reaction that generates heat.

Before the arrival of gunpowder, history is rich with examples of anecdotes of incendiary compositions being used. It is well-known that incendiary compositions were used in Assyrian siege campaigns and that the zikkam in the Old Testament may have been 'flaming arrows'. Athens was captured by Xerxes using fire-tipped arrows in 480 BC, and so on. An incendiary composition called Greek Fire was invented in Byzantium in circa 675 AD by a Jewish architect and Sryian refugee, Kallinikos. This was a primitive form of napalm that arguably contained saltpetre or *sal petrae* (meaning "salt of stones"). However, it was the invention of gunpowder, namely a mixture of saltpetre (potassium nitrate), sulphur and charcoal that changed the nature of warfare. The optimal formula for useable gunpowder is seventy five percent by mass of saltpetre, fifteen percent of sulphur and ten percent of charcoal. There had been many attempts to produce gunpowder with differing recipes resulting in various degrees of success in forming a useable propellant. Much of the gunpowder that was used in the early middle ages was weak and unpredictable in terms of its performance. The people who mixed it did so through trial and error and using their intuition knowing that small changes in procedure would result in a very different outcome. Preparing gunpowder could be a perilous task and those who sought fame and fortune through their technical prowess could be handicapped by a disfiguring burn, or even death.

---

[3]This is gravitational potential energy. Of course, other potential energies can be described and a more general definition would be that it is the energy derived from the object's position in a field, whether gravitational, electrical, magnetic or any other type of field (except a footy field!).

The origin of gunpowder is somewhat murky however the most authoritative modern view is that gunpowder first originated in the middle of the 9th C AD by Thang alchemists who were actually seeking for the elixir of immortality (Brown 1998). Early Chinese literature refers to a 'fire chemical' and 'fire drug' however these were probably used purely for their explosive effect and not as a propellant of a gun per se. Certainly by 1044 AD it was clear that the Chinese played with mixtures of saltpetre, sulphur and charcoal, usually with other ingredients such as oils, vegetable matter and arsenic compounds. The name *huo* yao was used to describe such mixtures. Partington describes these mixtures as proto-gunpowder as they were used principally in bombs and not as propellants (Partington 1960). True gunpowder came along in the latter part of the Mongol Yuan dynasty (1260–1368 AD) at a similar time to the discoveries in Europe. Certainly, by the 13th C the Mongols were using some form of propellant to launch fire arrows from bamboo, wooden and metal tubes. However, there may be some evidence that the Chinese were playing with the ingredients of gunpowder much earlier than that. A Chinese alchemical text from 492 AD noted that saltpetre burned with a purple flame thus providing a practical and reliable way of distinguishing it from other inorganic salts (Chase 2003).

The first hand-held firearm can be traced back to no later than the year of 1288 AD and was found in Manchuria in 1970. It consisted of a barrel that was 175 mm long and 25 mm in diameter. There was a chamber for the gunpowder that was 66 mm in diameter. The overall length including the chamber for the gunpowder would have been 340 mm and the mass was 3.5 kg. It would have been mounted on a long wooden stick—presumably to keep the person who wielded the weapon as far from the noisy smelly recoiling gun as possible. By the 1400s it seemed that the Chinese had moved into mass production of hand weapons. A popular model was the "heaven" (translated from the Chinese to English) presumably so named as this was the place that the user intended to send their enemy by ignition of the gunpowder! The average calibre was just over 15 mm. Each of these weapons was inscribed with the month and year of manufacture and oddly with a unique serial number, strangely foreshadowing modern firearm control. Thus we know that by 1436 there were 100,000 of these guns that had been made (Chase 2003).

In Europe, it is difficult to specifically identify when the first gun was fired however from the late fifteenth century a story has been in circulation that gunpowder and cannon were invented by a mysterious (and possibly fictitious) alchemist by the name of Berthold Schwartz, also known as Black Berthold (Partington 1960). It is not quite clear when Berthold allegedly made his invention although dates range from 1250 to 1353. However, it should be noted that Roger Bacon (1220–1292), a Franciscan Monk, had described gunpowder in ca. 1260. According to Partington, the legend goes that Berthold was testing Aristotle's theory that hot and cold natures were not to be mixed and were naturally antagonistic. Using a stone mortar as a vessel he mixed saltpetre derived from the earth (cold) with sulphur, (hot), together with some charcoal or linseed oil. The mortar was then put over the fire and the result was that it exploded, thus scattering bits of stone. Berthold was thought to be a necromancer and the theory that gunpowder was the work of the devil persisted well into the 17th

C. Erasmus (1466–1536) described the gun as an "engine of hell" and exclaimed: "Who can believe that guns are the invention of men?"

As for cannons, Partington (Partington 1960) reports of a cast iron cannon in existence dating from 1356 however a cast bronze cannon dating from 1332 is on display at the Beijing Museum of Natural History—see Fig. 1.1. Other cannons were appearing around Europe at a similar time and arguably even earlier: Seville is said to have been defended in 1247 by "cannon throwing stones" (Greener 1910). The first image of a gun can be traced back to 1326 in the manuscript authored by Walter de Milemete called *De Notabilitatibus, Sapientis, et Prudentia* that shows a gunner lighting the fuse of a what looks like a vase-shaped cannon mounted on a rather rickety table and pointing at a door (Partington 1960)—see Fig. 1.2. This is sometimes referred to as the 'Milemete Gun'. By all accounts, the Royal Armouries in the UK constructed a working model of the Milemete Gun using the gunner as a scale. The result was a very heavy gun weighing some 410 kg with a bore of 38 mm (Davies et al. 2019). Here the concept of these medieval weapons was simple in that the force of the gunpowder was being directed along the core of a cylinder enclosed at one end and this approach was still used well into the 19th C.

History records that the Guildhall, London had six brass guns for powder and lead shot as early as 1339 (Johnson 1991). The Milemete guns were made from bronze. If the gun was manufactured with a linear bore, that is a cylinder existed along the full length of the gun, then the bowing towards the breech end must have been due to reinforcement and not for the accommodation of a large quantity of gunpowder. Reinforcement at the breech showed some understanding of the pressures formed and the resulting stresses that would have arisen. One can only assume that this design was arrived at by trial and error and history does not record the gunners that lost their limbs, or even their lives during the cannon's invention! That is, unless you were a person of note. History does tell of the fate of the luckless James II of Scotland

**Fig. 1.1** A Bronze cannon from the Beijing Museum of Natural History discovered at the Yunja temple, Fangshan, Beijing in 1935. The four-character inscription on the body of the cannon reads" 3rd year Zhishan Era" which was in 1332 AD. *Source* Author

**Fig. 1.2** Earliest picture of a European cannon, Walter de Milemete, 1326

(1430–1460), who during the siege of Roxburgh Castle had his femur *"dung in two by a piece of mis-framed gune that brake in the shuting, by which he was stricken to the ground and died hastilie"* (Cleator 1967). Vase-shaped weapons appeared to be very popular at that time. A small vase-shaped cannon was also discovered in Sweden in 1861 and dated to a similar period as the Milemete gun—being tentatively dated to ca. 1340. In addition, another vase-shaped gun from Manuta in Northern Italy was carefully analyzed in 1786. This has been dated to ca. 1322 although it is now unfortunately lost (Davies et al. 2019).

## 1.3 Proliferation

By 1620, the great English scholar and the man behind the modern scientific method, Francis Bacon (1561–1626), who was no relation to Roger, noted that gunpowder was one of the three inventions that distinguished medieval times from ancient times. He noted that the other two notable inventions were the printing press and the magnet[4]. And, for good reason! Gunpowder had been used extensively in thousands of conflicts since its inception. During the Middle Ages, the battles weighed heavily on a man's muscular strength and the energy contained therein. However, with the invention of gunpowder and the gun there was a great leveller. It was certainly used in the battle between the English and the French in the battle of Crecy in 1346 and possibly even earlier. In this battle, the English used a device called a ribauldequin or 'organ gun', so called because of how it resembled a pipe organ. Multiple barrels were laid out in a line and fired simultaneously from a common fuse. The English employed twelve barrels which were used as a ranged anti-personnel weapon and had the capacity

---

[4]It should be noted that technically the Ancient Greeks were well aware of magnets and so they were not as modern as Francis Bacon originally thought!

to propel iron projectiles. It was successful in that it was one of the first weapons that enabled peasants to be proficient at killing the armoured French knights. No longer did a soldier have to wield a heavy sword to inflict injury onto his foe where success was largely governed by a person's height and strength. Now with the help of a long metal tube and gunpowder, even the weakest of soldiers could inflict a heavy injury on his enemy. In some ways it made war more accessible. Firearms allowed for a distinct level of remoteness from the enemy that only well-trained archers had previously enjoyed. Medieval archers were trained from a very young age to allow for the development of extraordinarily strong 'archery muscles' of the shoulder and upper back as well as the biceps and triceps. No longer was there a requirement to develop such muscles and still be some distance from your enemy to inflict injury. This remoteness made killing less risky, less messy and less tiring. Hand-to-hand combat was exhausting whereas pulling a loading and firing a gun was not.

It is because of its lethal nature, gunpowder soon became desirable to kings. In 1346 Edward III ordered all available saltpetre and sulphur to be brought to him for storage. In the first year of Richard II, the king ordered sulphur, saltpetre and charcoal to be sent to the castle of Brest. In 1414, Henry V ordered that no gunpowder should be taken out of the kingdom (Dillon 2015). The use of gunpowder became more prolific to the extent that by during the Queen Elizabeth's War with Spain, England consumed around 200 tonnes of gunpowder annually. By the middle of the 19th C. it was thought that a major war would consume 9000 tonnes annually.

Out of the constituents of gunpowder, sulphur and charcoal were relatively easy to come by. Afterall, sulphur is the tenth most abundant element in the Universe. When *Voyager I* passed by Jupiter's volcanic moon, Io in 1979, the surface was seen to be heavy in sulphur content (Soderblom et al. 1980). However, this was not much help for making gunpowder! The Earth on the other hand also has an abundant supply of sulphur. Sulphur had been known about for millennia with the earliest reference probably dating back 1550 BC for an Egyptian eye salve. During Roman times, sulphur was mined from the Greek Island of Melos and Cyprus. In medieval times, the market for gunpowder grew and so mines opened up across Europe, in Bohemia, Cracow (Poland), Italy and Spain and the Hekla volcano in Iceland. Other mines also appeared in Israel (Judea), Taiwan, India and Japan. It soon became a controlled substance with the Papacy imposing strict rules on prohibiting its export to non-Christian nations. In 1527, Pope Clement VII (1478–1534) issued a Papal order excommunicating those who sold sulphur to the Saracens. Similar decrees were issued by the Pope Paul III (1468–1549) and Pope Urban VIII (1568–1644) (Kutney 2007).

Charcoal was much easier to come by and had been used by metal workers for centuries to strengthen iron. The simplest way of mass-producing charcoal was to place a tree-like pole into the ground and surround it with a wigwam-arrangement of logs. The wood pile was then subsequently covered by earth and set alight from the top. This would have been allowed to smoulder for several days, after which the earth was removed to harvest the resulting charcoal debris.

Of the three constituents of gunpowder, saltpetre is the most difficult to harvest in large quantities. Satlpetre is what we call a 'metal nitrate', more specifically, potassium nitrate. It was soon discovered that urine, of all things, was useful in making saltpetre. Microorganisms in the earth turn urea, which is a compound found in urine and formed in the liver, into ammonia. Ammonia is a really useful molecule for explosive products—and still used today.

It should be noted too that soil is a highly underrated commodity. It is actually very important for sustaining our modern lives and is rich with life. It is thought that 1 gramme of soil from your garden will contain as much as 50,000 different species of microscopic organisms. Even in as little as one teaspoon of soil, it is thought there are more microorganisms than there are people on the planet.

Rather usefully for the medieval period, large domesticated animals like horses and cows produced heaps of urine. This would have seeped into the soil beneath their feet ready to be received by the microorganisms. The little urine-digesting organisms went on to combine the ammonia with oxygen to form nitrate ions. These nitrate ions would then combine with metals in the soil, namely magnesium, calcium and potassium to form a metal nitrate, the most important of which was potassium nitrate. Sometimes the dirt became so saturated with metal nitrates that crystals of these substances grew beneath floorboards or along the walls of basements.

In the fourteenth Century, saltpetre plantations became common place. The ingredients for the plantations were black earth (faecal material), harvested urine, dung, quicklime and occasionally oyster shells. Although the latter were more difficult to come by. The black earth, dung and oyster shells were used as a bed for the microorganisms, the urine was for the urea content that could be converted to ammonia and the quicklime was added to reduce the amount of acidity.

Such was the importance of urine for making the gunpowder that in 1626, it was proposed to Charles I that the beggars in the City of London should be employed to collect the city's urine for the purpose of making saltpetre. 'Saltpetre men' were employed by the Crown to recover saltpetre, or its active ingredients, and were known to dig up graveyards, invade the bed chambers of the heavily pregnant, the sick or the dyeing. It was quite a political bombshell at the time (Dillon 2015). The ideal donor was somebody who drunk wine or strong beer—ideally an alcoholic. This had to be the only time in history when being an alcoholic was actually useful for warfare! Of course, the bio-chemical reasons at the time were not known however it has come to light more recently that alcohol increases urea synthesis when the body metabolizes alcohol.

Interestingly, this was not the only military use of urine. Urea can also be used to manufacture a high explosive, urea nitrate, which has been used by terrorists and insurgents to produce improvised explosive devices. It was used in the 1993 attack of the World Trade Center—kind of a 'pee bomb', if you will. Further, polyurea, which is a compound containing repeated urea molecular structures, has been proposed to improve the blast protection of buildings due to its remarkable rubbery properties. One would have never of thought how powerful your urine could be! I digress.

Gunpowder's success was that a huge amount of gaseous oxygen was required to produce even small amounts of potassium nitrate. This oxygen was 'locked up', if you will, and able to be released through a chemical reaction. The basic principle of the chemical reaction was that the fuel (charcoal + sulphur) was mixed with the oxidiser, potassium nitrate. The oxygen atoms combined with the carbon and sulphur atoms and in doing so released a huge amount of heat energy and rapidly expanding gasses that can be used to push against the base of a projectile.

By the 19th C saltpetre refineries became common place to purify the saltpetre by separating it out from other metal-nitrates and impurities that would diminish the effectiveness of the final gunpowder product. In 1861, Major G W Rains from the Corps of Artillery and Ordnance noted, in his publication 'Notes on Making Saltpetre from the Earth of the Caves" that the government had established a refinery in Nashville, Tennessee capable of purifying 5000 lbs (2268 kg) of saltpetre per day for the purposes of manufacturing gunpowder (Rains 1861). This saltpetre had been produced by virtue of bat droppings and was known as *grough saltpetre*. Contemporary to Rain's publication, Joseph LeConte who was professor of Chemistry and Geology at South Carolina College, published his 1862 guidelines on the instructions for manufacturing saltpetre (LeConte 1862). This was a pamphlet that was written for "those who may be inclined to engage in the production of saltpetre". The idea was that the public would afford themselves of mining saltpetre to feed the insatiable desire for gunpowder by the State. The State would then refine it, after paying a reasonable price for the product.

To make gunpowder, the three important constituents were individually ground down, usually with a mortar and pestle and mixed together according to the appropriate formula. The result was "serpentine[5] powder". The name "serpentine" was probably a reference to the serpent in the Garden of Eden who deceived Eve. This mixture was deceptive! Due to the individual processes that were used to mill the individual constituents (saltpetre, sulphur, and charcoal), the powder could burn in an erratic way. Furthermore, a variable particle size would make transporting difficult. Small particles separate out from larger particles in a process called 'granular convection'. This is a process whereby larger particles will gradually move to the top of a collection of smaller particles when the collection is subjected to vibration. It is this annoying principle that means you must check your lawn for pebbles that migrate to the surface before trashing your lawnmower blade. Also, have you ever noticed why a bag of muesli tends to have the larger brazil nuts at the top and the smaller oat particles at the bottom? Well, it is the same effect (and interestingly, sometimes referred to as the 'brazil-nut effect' or the 'muesli effect'). This is not conducive to an even burn. So, a new method was required to make gunpowder.

It was soon realised that mixing the ingredients in a wet state, by adding water, was both safer and produced a more reliable manufacturing route. Once the serpentine powder was mixed into a slurry, it was allowed to dry as a sheet. Once dry, this "sheet" was broken down into small grains by hammers. The grains were then tumbled together to remove sharp edges and then passed through mesh screens for sorting into appropriate sizes. The grain size of the largest grains was typically about the size of a grain of corn (which is why the powder got the name "corned powder"). This was the powder that was most suitable for artillery pieces due to its size. Through the

---

[5]The lever mechanism use on matchlock guns was shaped like a serpent and also referred to as the serpentin, or serpentine matchlock mechanism.

centuries there were different methodologies employed to produce the same material all with different innovations to improve the quantity that could be manufactured, the consistency of the powder and the safety of manufacture. Still, it was a dangerous game and often there were accidents that resulted in the deaths of many workers.

Prior to 1846 (when nitrocellulose was discovered that paved the way for modern propellants) propelling projectiles was only achieved with gunpowder. However, gunpowder was not always reliably initiated, and it often left considerable residue. In ca. 1800 an early attempt was even made to see if mercury fulminate, a common explosive material, could be used as a substitute for gunpowder. Accordingly, Mr Howard FRS, bravely replaced 68 grains of gunpowder with 34 grains of mercury fulminate powder to see if it could efficiently propel a bullet. Alas the gun was split wide open and thankfully his accomplice was not hurt but a lesson was well learned: that certain explosive materials were not ideal for accelerating projectiles along a gun barrel!

Mr Howard wrote after his team's first attempt with 17 grains of mercurial powder and a lead bullet (Howard 1800): *"We therefore recharged the gun with 34 grains of the mercurial powder; and, as the great strength of the piece removed any apprehension of danger, Mr. Keir fired it from his shoulder, aiming at the same block of wood. The report was like the first, sharp, but not louder than might have been expected from a charge of gunpowder. Fortunately, Mr. Keir was not hurt, but the gun was burst in an extraordinary manner. The breech was what is called a patent one, of the best forged iron, consisting of a chamber 0.4 of an inch thick all round, and 0.4 of an inch in calibre; it was torn open and flawed in many directions, and the gold touch-hole driven out. The barrel, into which the breech was screwed was 0.5 of an inch thick; it was split by a single crack three inches long..."*

Finally, during the 1880s, a French inventor by the name of Paul Vielle (1854–1934) developed a nitrocellulose-powder which unlike previous powders, created little smoke or residue. Vielle revealed this synthetic "smokeless" powder, known as *Poudre B*, to the world in 1886. This was a new breed of powder and was about to displace gunpowder from its near millennia-old niche as the go-to propellant (Kelly 2004). Henceforth, the term "gunpowder" was used to describe the new synthetic formulation whereas "black powder" became synonymous with the ancient saltpetre-based propellant.

To know more about the manufacture of black powder the US Army have, very kindly, published an *Improvised Munitions Handbook* (Anon 2012). In it we read about improvised black powder where the description reads that *"black powder can be prepared in a simple, safe manner. It may be used as a blasting or gun powder"*. The list of materials required are of course potassium nitrate, wood charcoal, sulphur, alcohol, water, a heat source, two buckets (one heat proof), a flat window screening (such as a fly screen), a large wooden stick and a cloth. The recipe is listed as follows:

1. Place the alcohol in one of the buckets.
2. Place the potassium nitrate, charcoal, and sulphur in the heat resistant bucket. Add 1 cup of water and mix thoroughly with the wooden stick until all the ingredients are dissolved.

3. Add the remaining water to the mixture.
4. Place the bucket on a heat source and stir until small bubbles begin to form.
5. Remove the bucket from the heat and pour the mixture into the alcohol whilst stirring vigorously; let the alcohol mixture stand for about 5 min and then strain the mixture through the cloth to obtain the black powder.
6. Discard the liquid and wrap the cloth around the black powder and squeeze to remove all excess liquid.
7. Place the screen over the dry bucket and place workable amounts of damp powder on the screen; granules are produced by rubbing the solid mixture through the screen and then allowing to dry.
8. Hey presto! We have granulated black powder!

## 1.4 The Naval Revolution

During mediaeval times, it was unlikely that the presence of a gun on the battlefield had much effect on the outcome of the battle. In fact, in all likelihood the use of such primitive firearms was more of a risk to the user than the enemy and accidents occurred with startling frequency. However, around ca. 1350, guns found increasing employment in naval operations. Use on land sieges became less popular as the siege cannons were a logistical nightmare. For example, there was a monster cannon that was used in the of siege of Constantinople and it was reported that this had to be dragged by 30 linked-wagons drawn by team of 60 oxen with 450 workman, 250 of whom were employed in preparing the route. As you can imagine, progress was slow. And they were able to progress at a rate of three miles per day and so two months were spent in travelling a distance of 150 miles. Ironically, this was a lesson that Hitler did not learn in the Second World War with his huge railway guns. More on that later. The ridiculous size of some siege cannons lead Charles VIII of France (1483–1498) to redesign a cannon such that it was limited in length to eight feet, or 2.44 m (Cleator 1967). Further he also insisted on trunnions being fitted to the barrel which allowed for elevation and depression of the gun in an easy fashion. These guns could also be mounted onto carriages which could be pulled by fast and fit horses. And thus, the field gun was born. Further advances in these types of guns were aided by the discovery of how to melt iron. No longer bronze had to be used but instead a more resilient and stronger metal in iron could be used to manufacture the gun barrels. Knowledge of this accomplishment spread quickly.

Warships of course do not suffer from this logistical impediment. Warships could accommodate quite sizeable cannons, some of which could weigh up to four tonnes (see Table 1.1). These were placed on a wooden carriage that would accommodate the gun recoiling.

Into the sixteenth century it was commonplace for war vessels to employ cannon. The Mary Rose was one such ship that famously sank during the reign of Henry VIII in around 1545. The Mary Rose was standing in the Solent, off the coast of Britain. During a distant firing from a French warship she was sank together with her commander and 600 men. Several theories exist for the reason why she sank. One strong possibility is that she was overloaded by the weight of her cannons and crew coupled with an eye-witness account that there was a strong gust of wind around the time of the sinking. Perhaps the overloading and the strong gust of wind worked harmoniously as she opened her lower gunports thereby allowing the seawater to flood into the toppling ship. Or, was it the French guns? Perhaps, we will never know for sure. Notably some of the cannons that were used on the Mary Rose were of a breech-loading type which were not to become mainstream until around the nineteenth century.

The guns used on the Mary Rose comprised of a tube of iron, made from staves over which successive hoops of iron were driven whilst being red hot. On cooling, these hoops compressed the inner iron tube leading to a product that accommodated larger quantities of propellant than would have been otherwise supposed. In effect, the addition of the hoops allowed for a more even distribution of stresses that would arise during firing. No one knows how the early gun designers understood that the use of external hoops improved the performance of the gun in terms of being able to accommodate larger propellant masses. Again, one must assume that they discovered this through a process of trial and error. The gun was affixed to the large timber beam and loading of the gun occurred by simply removing a breech block, inserting a projectile and charge, replacing the block and wedging it into the barrel from behind (Greener 1910). Ropes were attached to aid in preventing excessive gun jump and recoil during firing. In addition, a "bitt", or large beam was fixed perpendicular in the deck to limit recoil.

By the time Elizabeth I ascended to the throne in England there was a developing taxonomy for guns deployed on the Armada. Every one of these guns were heavy beasts and, of course, susceptible to recoil when fired resulting in a stark health-and-safety hazard as they rolled back on the decks! According to one experiment with an unrestrained 32-pounder weighing 3.5 tons (gun plus carriage), the gun recoiled in excess of 15 m on firing. To prevent the gun from disappearing through the other side of the ship, heavy breeching ropes secured the gun to the ship's structure with a pair of ring bolts. The force on these ring bolts was enormous. If a simultaneous broadside was fired, there was the potential for damage to the ship. By one estimate,

**Table 1.1** Types of Armada guns from (Lewis 1961)

| Name | Typical Calibre, in | Typical mass (in 1595), kg | Typical barrel length in calibres | Typical shot mass, lb, (kg) |
|---|---|---|---|---|
| Cannon | 7.25 | 3864 | 18 | 50, (22.7) |
| Demi-cannon | 6.25 | 2045 | 20–22 | 32, (14.5) |
| Cannon perier | 8 or 6 | 1295 | 8 | 24, (10.9) |
| Culverin | 5.25 | 1818 | 18–32 | 17, (7.7) |
| Demi-culverin | 4.25 | 1295 | 32 | 9, (4.1) |
| Saker | 3.5 | 818 | 32 | 5, (2.3) |
| Minion | 3.25 | 545 | Possibly 32 | 4, (1.8) |

if HMS Victory fired a single broadside with all the guns firing together, the force on the hull-frame would have been equivalent to the weight of 1300 tons. Hence, the practice of firing in a ripple from head to stern was developed (Goodwin 2016).

A list of 'typical' weaponry deployed on the ships at that time is presented in Table 1.1 (Lewis 1961).

### 1.4.1   Turret Development

By the nineteenth Century it was notable that the guns used by the Royal Navy had not changed since Napoleonic times. The muzzle-loaded smooth-bore cannons were notoriously inaccurate. Evidence for this was the fact that very few ships were ever sunk directly by gunfire. In October 1805 Nelson's fleet attacked the combined Spanish and French fleets off Trafalgar. This was the largest sea battle that had ever been fought and indeed has since. The battle lasted for over six hours and not one British ship had been sunk by enemy gun fire. For success, proximity was required. However, at the right distance, a broadside attack could be devastating. The mass of iron that could be discharged from a single broadside attack from HMS Victory was 522 kg. Victory's first opening broadside at Trafalgar, fired into the stern of the French 80-gun flagship, Bucentaure, passed right through the ship and killed 197 men and injured 85 including the Captain, Jean-Jacques Majendie (1766–1835). This is perhaps unsurprising as the estimated muzzle velocity of the projectiles from Victory's carriage guns was close to 490 m/s and therefore supersonic—like many weapon systems of today (Goodwin 2016).

A solution to the problem of limited accuracy was to introduce rifling. This also involved the redesign of a shell so that it was no longer spherical but now 'bullet shaped'. Parallel to this, ship armour was evolving. The French Gloire (=Glory) was launched in 1859 which was one of the first iron-clad armoured ships. The Gloire boasted 12 cm-thick (4.7 in) armour plates, backed with 43 cm (17 in) of timber. In competition with the French, in 1860 HMS Warrior was launched with 10 cm (4 in) of iron backed by 46 cm (18 in) of teak. This was the largest and most powerful of any iron-clad warship afloat (and is still afloat in Portsmouth, UK). It had a compliment of over 700 officers and men yet still had broad-side gun decks and smooth-bore muzzle-loaded cannons.

Unfortunately for the British and French, hundreds of years of tradition were about to be abandoned when the Swedish designer John Ericsson developed the first ship with a rotating turret. Turrets would revolutionise naval warfare. They would reduce the need for a large number of static guns that were henceforth required to cover the perimeter of the vessel. It meant that the ship could be more versatile in its design and instead of 40 medium-sized guns, as was the case with the HMS Warrior, it was possible to install two large and powerful guns that could cover all azimuths. For example, the USS Monitor, which by and large was regarded at the time as foolish, was an iron clad with a low-profile raft-like structure and a central turret. Ericsson understood the very basic principles of survivability, that many Naval architects of the time did not, that a small target was relatively difficult to see and hit (Thompson 1990).

The USS Monitor was designed with a single requirement in mind: to defeat the CSS Virginia. The Virginia had a large deck with a powerful battery that could overwhelm most ships. She also carried a thick armour belt that was angled so that light shot would simply be deflected. So, the Monitor had to be fitted with suitably large guns. On January 30, 1862 Ericsson was told that the Monitor would inherit the two standard length Dahlgren guns from the USS Dacotah. These were guns that had been so designed with a smooth profile and a large thickness of steel to accommodate the pressure in the breech at the chamber end of the gun.

The Monitor fought battles in the American Civil War but most notably confronting the Virginia on Sunday 9 March 1862, achieving what she was built to do. Subsequently, Ericsson was paid his reservation fee that had been held back by the United States Navy (Thompson 1990). Ericsson lived to the age of 86 and although he died in America his body was interned in his native Sweden presumably because of his success with turrets.

# References

Anon (2012) U.S. Army improvised munitions handbook. Military tactics. Skyhorse Publishing, New York, USA

Brown GI (1998) The big bang: a history of explosives. Sutton Publishing Ltd, Gloucestershire, England

Chase K (2003) Firearms: a global history to 1700. Cambridge University Press, Cambridge, UK

Cleator PE (1967) Weapons of war. Robert Hale, London, UK

Davies J, Shumate J, Hook A, Walsh S (2019) The medieval Cannon 1326–1494. New Vanguard. Bloomsbury Publishing, Oxford, UK

Dillon B (2015) The great explosion: gunpowder, the great war, and a disaster on the kent marshes. Penguin Random House UK, UK

Eigenbrod L, Chevedden PE, Foley V, Soedel W (1995) The trebuchet. Sci Am 273(1):66–71. https://doi.org/10.1038/scientificamerican0795-66

Goodwin P (2016) The practice and power of firing broadsides in british men of war during the age of fighting sail. Arms Armour 13(1):48–61. https://doi.org/10.1080/17416124.2016.1191750

Greener WW (1910) The gun and its development, 9th edn. Bonanza Books, New York, USA

Howard E (1800) XI. On a new fulminating mercury. Philos Trans Royal Soc London 90:204–238. https://doi.org/10.1098/rstl.1800.0012

Johnson W (1991) Some monster guns and unconventional variations. Int J Impact Eng 11(3):401–439. https://doi.org/10.1016/0734-743X(91)90046-I

Kelly J (2004) Gunpowder: alchemy, bombards, and pyrotechnic. Basic Books, New York, USA

Kutney G (2007) Sulfur: history, technology, applications and industry. ChemTec Pub, Toronto

LeConte J (1862) Instructions for the manufacture of saltpetre. South Carolina College, Columbia, South Carolina

Lewis M (1961) Armada guns: a comparative study of English and Spanish armaments. Allen & Unwin

Partington JR (1960) The history of greek fire and gunpowder. W. Heffer & Sons Ltd, Cambridge, England

Rains GW (1861) Notes on making saltpetre from the earth of the caves. Corps of Artillery and Ordnance, New Orleans, USA

Soderblom L, Johnson T, Morrison D, Danielson E, Smith B, Veverka J, Cook A, Sagan C, Kupferman P, Pieri D, Mosher J, Avis C, Gradie J, Clancy T (1980) Spectrophotometry of IO: preliminary voyager 1 results. Geophys Res Lett 7(11):963–966. https://doi.org/10.1029/GL007i011p00963

Thompson SC (1990) The design and construction of USS monitor. Warship Int 27(3):222–242

Wheelis M (2002) Biological warfare at the 1346 siege of Caffa. Emerg Infect Dis 8(9):971–975. https://doi.org/10.3201/eid0809.010536

Williamson GA (1976) Josephus—the Jewish war (translated with an introduction by Williamson GA). Penguin Books Ltd, Harmondsworth, Middlesex, England

# Chapter 2
# How Guns Work

Guns have not changed too dramatically over the centuries. However, there were several steps that guns went through to reach what we understand as a modern gun system. We will now see how important changes have occurred and examine the fundamentals of a gun's working.

## 2.1 Principia

The behaviour of a gun during firing is subject to the same known laws of the universe just like everything else. And it is these laws that dictate its behaviour and whether the projectile it fires is able to deliver the appropriate target effects.

In 1686 Isaac Newton published his three laws of motion in "Principia Mathematica Philosophiae Naturalis". Of course, as we have seen, guns already existed at that time. Nevertheless, these three laws tell us a lot about why guns function the way they do. Let us look at these laws and find out why they are so important to gun owners.

### 2.1.1 Newton's First Law of Motion

So, let us start with what is commonly called: Newton's first law.

> An object at rest will remain at rest unless acted on by an external force. In addition, an object in motion continues in motion with the same speed and in the same direction unless acted upon by an external force.

This law is often called "the law of inertia" and this is very important when we consider the trajectory of projectiles fired from artillery guns and the like. All projectiles are subject to the laws of gravity. Without gravity and air resistance, a gun

© Springer Nature Switzerland AG 2021
P. J. Hazell, *The Story of the Gun*, Springer Praxis Books,
https://doi.org/10.1007/978-3-030-73652-1_2

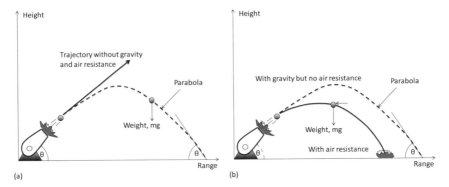

**Fig. 2.1  a** Trajectory of a gun-fired projectile without gravity and air resistance (solid line) and the trajectory of a projectile with gravity but without air resistance (dashed line) and **b** the trajectory of the projectile subjected to both air resistance and gravity (solid line). *Source* Author

elevated at 45° could fire a projectile that would carry on a straight trajectory for an infinite distance, or until the projectile hit something. OK, so let us introduce gravity. The projectile, is now under the downward pull of gravity and will follow a perfect parabola (that is, a uniform and symmetrical curve) such that it hits the earth at an angle that is identical to the angle at which it is launched (e.g., 45°—see Fig. 2.1). The next step is to introduce air resistance. Projectiles experience enormous drag as they travel through the air and the reason is that air has mass. It has density. So, being mostly nitrogen, with a good helping of oxygen (thankfully) and a smidge of argon, carbon dioxide, neon, helium and methane, it has a density of 1.225 kg/m$^3$. That is around 1/800th of the mass of water for the same volume.

The air resistance is a particular nuisance to projectiles that are travelling at speeds that exceed the speed of sound in air (342 m/s at sea level). And large drag forces are imparted to the projectile as we approach what we call the trans-sonic region (around about the point where the speed of sound is broken) and the supersonic region. Most high-velocity rifle bullets, artillery shells, and tank-gun projectiles travel at supersonic velocities and therefore are subjected to the highest drag forces.

Historically, it was largely thought that projectiles would fly through the air and abruptly come to a sudden stop and so drop to the ground. Of course, that was before we had the benefit of high-speed visualisation equipment, or even Newton's science. Early philosophers pondered on this question: what happens to a cannonball once it has left the cannon? Even with the slowest of projectiles, it was not possible to see its passage through the air with the naked eye. So, it remained a mystery for hundreds of years.

It was an Italian mathematician by the name of Niccolo Tartaglia (1500–1557) that started to make strides to answer the questions that had deeply mystified philosophers and scientists for decades. He was asked once, 'what is the angle of the cannon required to achieve the longest range for a projectile?'. During his research he found that it was, indeed 45° (which is true in a vacuum) and so he proceeded to invent the gunner's square. This had one leg of a carpenter's square that fitted down the barrel of the gun with a pendulum attached that gave the angle of inclination. For the first time, it was apparent that accurate instrumentation had made its way into gunnery (Kelly 2004).

## 2.1.2  Newton's Second Law of Motion

Let us now move on to Newton's second law which states that:

> The rate of change of the momentum of a body with time is equal in both magnitude and direction to the force imposed on that body.

That sounds complicated, so let us simplify that somewhat. The momentum of a body is equal to the multiplication of its mass with its velocity. However, if we assume that the mass of the body is constant and does not change with time then we can separate out mass and velocity. Our consideration can then become the rate of change of velocity with time, which is acceleration. Newton's second law of motion can therefore be written as:

$$F = m.a$$

where $F$ represents the nett force acting on the body, $m$ is the mass of the body and $a$ is the acceleration. Both the force and the acceleration are *vector quantities* in that they both have size and direction. Mass is a *scalar quantity* in that it only has size.

In a gun we have a projectile that has mass, to which a nett force is applied by the expanding propellant gasses minus any friction that maybe present between the projectile wall or jacket and the gun barrel. This nett force results in projectile acceleration. Frictional forces on the projectile are usually quite small. Nevertheless, it is important to minimise any resistive forces acting on the projectile during launch. Newton's second law of motion also comes into play when the projectile strikes a target. The target, depending on its strength and other factors, will apply a massive resistive force on the projectile and therefore cause it to decelerate as it penetrates the target material. Harder, stronger target materials will cause the projectile to decelerate more sharply than softer materials as we will see in Chap. 8.

## 2.1.3  Newton's Third Law of Motion

Newton's third law of motion simply states that:

> For every action there is an equal and opposite reaction.

Take your finger and press it against your desk. You will be applying a force to the desk which will attempt to push it down and at the same time your finger will be compressed as you experience the reaction to that force. If we think of a gun system that fires a projectile, expanding propellant gases will cause the projectile to move in one direction due to the force that is being applied to it. Simultaneously there will be an equal and opposite reaction force due to the applied force. This results in the gun recoiling. A lovely example of this third law of motion in action was with the introduction of the Davis gun developed in 1910.

Commander Cleland Davis (1869–1948) of the US Navy developed his non-recoiling gun whilst serving on the USS Mississippi. The recoilless nature of the gun allowed it to be fitted to quite small supports. Its principal use was for mounting in the nose cockpit of a Curtiss NC-series flying boat. This was so that the flying boat would be able to shoot targets located on or just under the sea such as submerging submarines. The Davis recoilless gun comprised of essentially two long barrelled firearms which were fitted back to back. See Fig. 2.2. The forward-pointing barrel was rifled and carried a conventional projectile whereas the rearward-pointing barrel was smooth bore to carry the anti-recoil lead-shot. They were fitted such that the rear of one breach fitted to the rear of the other breach.

Arguably, Davis was not the first to stumble over this idea. In fact the great scientist and inventor Leonardo da Vinci (1452–1519) came up with a similar design many hundreds of years earlier (Olcer and Lévin 1976).

The ammunition was quite unusual. It comprised of a single cartridge case with a conventional shell and powder charge. Behind the powder charge located a barrier which separated the powder charge from the anti-recoil system comprising of lead shot, Vaseline and grease. The primer fed in from the side of the cartridge case and was located approximately halfway between the base of the shell and the lead

**Fig. 2.2** The Davis recoilless gun (Public Domain)

shot. To fire the weapon the barrels were separated, and the projectile was fitted to the forward pointing breach. On firing, the propellant charge was ignited, and the projectile accelerated down the forward pointing barrel. Simultaneously, the wax and lead shot combination was ejected out of the rearward barrel. Thus, any recoil was cancelled.

In some ways, it demonstrated that Davis had a good understanding of Newton's third law of motion. He knew that the rear-throne mass comprising of lead shot and wax had to weigh the same as that of the projectile. And so, the gun did not recoil but rather momentum was transferred to the rear-ejected mass of lead shot and wax (rather than the gun structure).

The life of this gun was short-lived as the system was quite cumbersome and clearly dangerous to overhead allied aircraft. However, the Germans readopted the principle in attempting to arm aircraft with 88 mm gun to attack surface targets. The recoil momentum due to firing the seven-kilogram projectile in the 88 mm weapon was balanced by accelerating the cartridge case of a similar mass in the opposite direction. The idea was that it was mounted under the aircraft fuselage and the intention was to dive attack against battleships and other important targets.

## 2.2 From Guns to Rockets

There are different methods for attacking an enemy target as shown in Fig. 2.3. With a gun we have the potential to fire a projectile to a high velocity and with

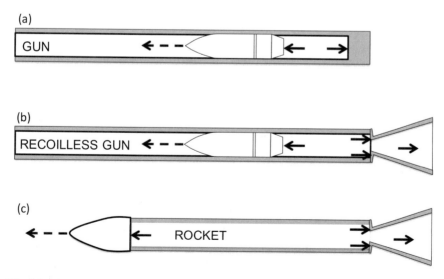

**Fig. 2.3** A schematic of different types of weapon systems **a** a simple gun, **b** a recoilless gun and **c** a rocket. All use propellant to produce thrust, but with different effect. *Source* Author

good accuracy, but the drawback is that you potentially have a heavy carriage for the payload that you deliver to the target (see Fig. 2.3a). If you shot lots of projectiles from a well-designed gun, then hypothetically you are going to get a low dispersion of hits. You will also need some form of gun mount and be able to control the recoil forces and so forth. Your range is limited too.

A recoilless gun as the name implies, is recoilless (see Fig. 2.3b). Modern recoilless guns are loosely based on the invention by Charles J. Cooke. In 1921 Cooke patented a system using vented propellant gasses to balance the recoil and it was his invention that formed the basis of the modern recoilless rifle (Olcer and Lévin 1976). Some forces will still be acting on the gun structure but most of the reaction is vented rearward through a nozzle to balance the force acting on the accelerating projectile. The advantage of a recoilless gun is that you can dispense with a heavy carriage. Having the lighter-weight carriage is advantageous because it means that you can put the gun on the back of pick-up trucks and jeep-like vehicles. In Australia there was a 105 mm recoilless rifle (M40) which was badged as a 106 mm system because there were other similar 105 mm guns and ammunition mix-up needed to be avoided. The disadvantage of these systems is that you need more propellant and of course you get a lot of dust kicked up at the rear of the weapon system. Range is limited too.

The next iteration of how we control the propellant burn is by the introduction of the rocket (Fig. 2.3c). Here the thrust is not applied to a moving projectile but is in fact applied to the whole body of the rocket and so the advantage here is that these provide for a long range and a light-weight launch structure. The disadvantage is that you require a higher propellant consumption. Arguably the other advantage here is that you are not so limited on the size of your payload or the range; you can launch very large rockets to large distances, and these can cause lots of devastation.

## 2.3   The Burn

A gun is actually a form of a heat engine. That is, like an internal combustion engine of a car, fuel + oxygen is ignited that drives a piston in a cylinder. Of course, with an internal combustion engine, that piston is connected to a con-rod which in turn drives a crank shaft. In a gun, the fuel (the propellant) contains its own oxygen (and therefore there is no need for an air-inlet) and the projectile is analogous to the piston. There is no con-rod and the cylinder is the barrel. To ignite the propellant a little sensitive primer cap is the starting mechanism, and this would ordinarily contain a sensitive explosive, that, once impacted would effectively spark and provide initiation for propellant burn. This was an approach that was first developed in ca. 1820 and allowed for the firings of guns in all weather. Before that it was necessary to use a 'flint-lock' device where a steel percussion lever would come into contact with a flint to cause a spark and initiate the gunpowder. This could be extremely tricky in wet conditions and the result was often quite variable. Primer caps contain materials such as mercury fulminate—a sensitive explosive that is quite sensitive to impact and heat. Mercury fulminate is a primary explosive.

The fact a gun is similar to an internal combustion engine was first stumbled upon by the Dutch theorist Christiaan Huygens (1629–1695) in 1673 (Kelly 2004). Huygens was a genius of his time and had a vision for using gunpowder for good rather than ill. The beauty of gunpowder, it was thought, was its possibility as a compact energy source. The key was to harness the explosive force. His *moteur a explosion* used the explosive force of gunpowder to lift a piston enclosed in a cylinder. Huygens used a vertically orientated cylinder, into the bottom of which was fed gunpowder. As the gunpowder was lit and burned, the hot gases accelerated a piston upwards. As the piston moved upwards it eventually uncovered some holes in the cylinder that allowed for the venting of excess propellant gases. The partial vacuum that resulted as the remaining propellant gases cooled, allowed for the piston to return. Thus, the gunpowder engine was born. By all accounts Huygen calculated that a single pound of gunpowder could lift a 3000-pound (1364 kg) mass through a distance of 30 feet (~ 9 metres). Alas, Huygens invention never came to anything due to the practical realities of delivering multiple strokes. Nevertheless, his work eventually was realised two centuries later with the invention of the internal combustion engine.

## 2.4 Let Us Talk About Recoil

Everyone who has ever fired a gun knows a little about recoil. In fact, a cursory search on YouTube reveals a number of hilarious videos of people unknowingly being caught out by one of the fundamental laws in the Universe: that for every action there is an equal and opposite reaction. Which as we have seen, was first realised (or articulated) by Newton. In fact, that this is the essence of recoil. Recoil can be very serious for gun operation and for larger guns the recoil force can be quite sizeable. In these cases, if the gun barrel is constrained, then large forces are transmitted to the structure and will cause the gun structure to fail or potentially flip. Recoil can also affect accuracy if the gun barrel is not allowed to travel rearwards in a linear motion. Thus, careful thought must be given to the design of a recoil system. The force due to gun recoil is applicable to a whole range of guns and not only from small-arms weapon systems but right through to fairly sizeable artillery systems and big direct fire weapons systems. It is the force that gave rise to one of my favourite quotes of all time from a Hollywood movie. In the most recent depiction of the A-Team John "Hannibal" Smith (Liam Nielson) and the team were plummeting to Earth in a tank. The only way of surviving the fall was to land in a lake. And so, Hannibal gave the orders to fire the main gun to use the force of recoil to fly the tank!

Flight Control Commander: Are they trying to shoot down the other drone?

Charissa Sosa: No, they're trying to fly that tank.

Now, actually the force of the gun would *strictly* be accommodated by the recoil system. However, the braking force applied by Hannibal's tank recoil system that brings the recoiling gun to rest is subsequently transferred to the whole system, thereby shifting the tank in flight. In 'real life', it is commonly seen by artillery crews as the dirt is kicked up behind the spades, or when a tank crew fires a shot causing the vehicle to slip.

As we have seen in Chap. 1, some early examples of guns that employed some form of recoil system were the cast-iron cannons of the 18th C. These would fire large cannon balls or shot from the cannon. The cannons were themselves mounted on wheels and therefore during firing the cannons would roll back often constrained by some heavy ropes. It was the friction between the rope and the steel eyes, through which the rope was fed, that led to the force of the firing being dissipated. Furthermore, rolling friction (that is the friction that is apparent due to the normal force from the reaction of the weight of the cannon distributed over the wheels) also acted to dissipate the force. Other cannons were mounted on inclined steel structures where the cannon would rise during the firing process as the barrel moved rearward (with the force of gravity pushing down). In both cases it is apparent that a small resistant force acts to push against the gun.

The way that engineers resolve these problems is to talk in terms of *impulse*. Impulse is calculated by multiplying the force applied by the time over which that force is applied. It is also equivalent to the change in momentum of a particular object (see Fig. 2.4).

As the gun fires, there are three masses that are being accelerated and gain momentum. The first is the projectile. Obviously, the main function of a gun is to fire a projectile toward its target. The second mass in question is the mass of the recoiling parts. This will include the gun barrel and indeed for a small-arms weapon-system would be inclusive of the whole weapon. The third mass is due to the propellant. That mass does not disappear into thin air. In fact, that mass also plays a role in recoil. For example, in a 5.56 mm × 45 mm cartridge there would be approximately 4 grams of propellant contained by the brass cartridge case. That 4 grams of propellant would need to be converted to 4 grams of gas; the mass is conserved as it is throughout the rest of the universe. For a 120 mm tank gun the mass of that propellant would weigh approximately 8 kg and that 8 kg of propellant gains momentum in the form of expanding propellant gases. The same applies for gas guns that use pressurised light gasses as the propelling medium. In my experience, students seem to have an intrinsic belief that gases such as hydrogen and helium do not weigh anything because they are lighter than air. The reality is that they have mass and therefore we need to account for that mass when examining the effects of recoil.

For large guns there is usually something called a 'buffer'. As the gun is allowed to recoil reward a hydraulic fluid is forced through orifices in a conical structure so as to provide a controlled rearward resistance during firing. Figure 2.5 shows a schematic of a typical buffer system. Of course, no gun uses a design like this and this has been drawn purely for illustrative purposes. For one thing, there is no sign of the conical structure that I mentioned just earlier! You will see that there is a hydraulic reservoir against which a piston pushes during the recoil action. So as the piston moves rightwards, hydraulic fluid is forced from the 'high-pressure' zone on

**Fig. 2.4** Graph showing the mechanisms of recoil. The gun delivers an impulse by virtue of the rapid propellant-burn that lasts for a short duration of time. This is counteracted by a relatively small force applied for a longer duration of time by the recoil system. *Source* Author

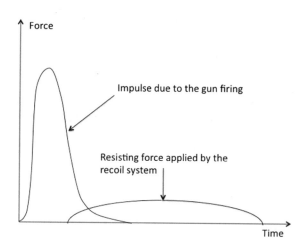

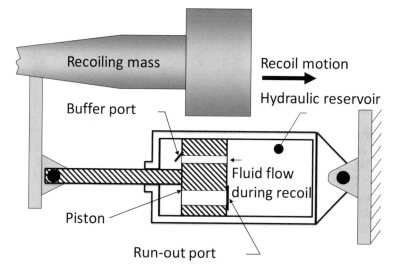

**Fig. 2.5** A schematic of a buffer system used for accommodating recoil during a shot. *Source* Author

the right, through an orifice in the piston, to the low-pressure region on the left. Thus, the speed at which the recoiling mass (that includes the mass of everything on the gun that moves rearward during firing) can be controlled by varying the size of the hole in the piston. And, the reaction force acting on the piston as it recoils is reduced by virtue of allowing the fluid to flow through the piston.

The next piece of the puzzle is the recuperator. This can merely be a powerful spring system comprising of multiple springs that are compressed when the gun recoils. Subsequently, they push the gun back to its firing position after the recoil motion has come to rest. The kinetic energy of the recoiling mass is transferred to stored elastic energy in the springs and this is what is used to bring the gun back to its firing position. And this needs to occur in a controlled fashion. So, next on the list we have something called the 'control-to-run-out' and this is usually a hydraulic dampening system that stops the metal parts crashing together from the stored energy in the recuperator.

The length of the recoil stroke is a very important parameter and for tank guns is of the order of 300 mm. For lightweight tanks this might be increased to 700–800 mm to accommodate a lower braking force and therefore less possibility for vehicle slip during firing. For something like an AR-15 assault rifle, a comfortable recoil would be of the order of tens of millimetres; for a 155 mm howitzer it would be more like 1500 mm.

Many people who fire guns will be familiar with the inevitable lift of the gun barrel when the gun is fired. This is simply due to a question of geometry as the shooter will hold the gun below the barrel along the axis of which, the recoil forces act. So, as the gun recoils there is a tendency for the hand to tilt upwards. Recoil can also cause serious injury.

The force imparted on the cartridge is governed by a number of factors and these include:

1. The mass of the weapon
2. The mass of the projectile
3. The mass of the propellant; and
4. The velocity of the projectile.

This may seem odd but the heavier the weapon, the lower the recoil force. This is due to the fact that it is much more difficult for the propellant gases acting on the base of the cartridge to accelerate the gun. However, the higher the velocity of the projectile, the mass of the projectile and the mass of the charge, the higher the recoil that is experienced by the shooter.

This was realised by the introduction of the infamous 'Nock Gun'. The Nock Gun was a seven-barrel shooter where six barrels were symmetrically arranged around a seventh barrel. This was kind of a copy of the multi-barrelled weapons of the 14th C. with the exception that these were designed to have specifically as a shoulder gun. On the pull of the trigger, seven projectiles were fired simultaneously. The trouble was, that the inventor had not quite got to grips with the conservation of momentum in that the momentum of the products moving forward must equal the momentum of the products (i.e. the gun) moving backward. It even became known as a 'shoulder breaker' as the force of the recoil was so great (Fig. 2.6).

### 2.4.1  Muzzle Brakes

Muzzle brakes are commonly fitted to guns to aid in reducing the recoil. The main purpose is the redirect the propellant gasses such that the amount of forward-moving inertia is reduced. They come in all shapes and sizes and it is most unusual to find

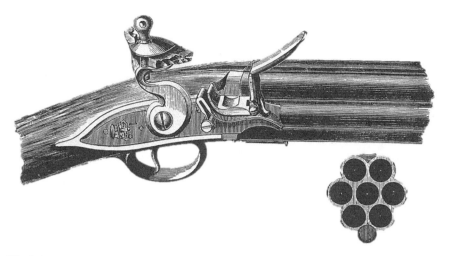

**Fig. 2.6** Nock's seven-barrelled carbine (Greener 1910)

them absent in larger calibre weapon systems and they are commonly used on small-arms weapon systems. Although there is a clear advantage in using a muzzle brake to reduce the recoil of the gun, muzzle brakes can have some disadvantages too. An installed muzzle brake can lead to a relatively high back pressure. This is called an overpressure that can have a deleterious effect on the health and comfort of those around the gun. We know that certain pressures can lead to injury so for example it is well known that the human tolerance for blast overpressures is about 2 psi and above this level the ear drums could rupture. Lung damage can occur anything between 20 to 30 psi so you would expect to see injury occur if there is a large back pressure due to the presence of the muzzle brake. This type of risk would only be associated with large artillery guns with efficient brakes. Occasionally a deflector fitted behind the brake can reduce that effect.

There are also other disadvantages of muzzle brakes in that they have their own finite mass and so there may be a requirement to add a counterbalance, especially for large guns. Muzzle brake will also have a negative effect on muzzle velocity (although this is usually quite small) as well as the vibration that would be experienced by the gun barrel during the firing process.

Two principal muzzle brake systems that are commonly used are:

- A pepper pot muzzle brake that are specifically designed for tank guns and it means that saboted sub-projectiles would be able to be launched from the gun.
- A baffled muzzle brake is generally used where the calibre of the projectile matches that of the boar of the gun. These are very commonly used in artillery guns.

The other possible drawback of using a muzzle brake is the fact that certainly in the direct fire use (i.e., tank guns), the presence of a muzzle brake can lead to obscuring the shooter's line of site to the target.

## 2.5   The Evolution of Rapid Fire and Automatic Operation

It is quite a remarkable achievement that many models of guns today have the ability to fire a bullet, eject a spent cartridge case and then reload and fire a subsequent projectile repeatably and reliably all in the timing of a 1000th of a second.

Rapid-fire weapons had always been the "Holy Grail" of weapon design and it was not until the eighteenth century that workable reliable models began to appear. The first 'revolver' type gun can be traced back to a patent in 1718 with the "Puckle Gun" invented by James Puckle (1667–1724). The tri-pod mounted gun involved a manually rotated drum magazine comprising of between 6 and 9 chambers. Each chamber was pre-loaded with black powder and projectile. As each chamber was manually lined up with the barrel, a cover was opened that exposed a flash-pan containing a small quantity of primer powder. This allowed a conventional flintlock mechanism to ignite the primer powder. The drum came with two options: with round chambers and square chambers. Oddly, this allowed the user to fire spherical projectiles if the target was a Christian, whereas cube-like projectiles if the enemy was a Turk! Although it was thought that cube-like projectiles would have been more painful for the enemy, I cannot help but feel that the instability of its travel along the barrel would have made it less accurate and slower than its spherical counterpart.

A trial of the weapon was recorded in the *London Journal* of 31 March 1722 that showed that the gun was capable of firing 63 projectiles in 7 min (Ffoulkes 1969). Sadly, the invention was not taken further and saw little commercial success. This was a pity given that it preceded Samuel Colt's hand revolver by 118 years.

Later, a more devastating rapid fire weapon was developed by Richard J Gatling (1818–1903), a middle-aged medical graduate. As a medic, he would have been aware of the number of men who would die in conflict purely from disease. If, he thought, you could replace a platoon of men with a single man who operated a weapon that can do the work of many weapons, then less men would have to fight and therefore less men would die. This assumption led to the development of a new weapon system: The Gatling gun. In 1862 Gatling submitted his first US patent for a rapid-fire battery gun which famously comprised of six rifled barrels that rotated about a central axis by the action of a manual rotating crank. As the barrels were rotated, ammunition was fed into the breach from above by a hopper. And so, the term *machine gun* was born. Gatling's invention was revolutionary in that for the first time, streams of bullets could be fired at up to 150 times per minute. However, his invention struggled to gain acceptance initially and was hindered by the poor metallurgy of the ammunition and the chemistry of the gunpowder that would cause jamming.

Technically, there is nothing new under the sun, as they say. In fact, early revolving cannons, some of large calibre, are described in mediaeval manuscripts. For example, there is one four-barrelled arquebus that was dated to the first half of the 16th C. and may have even belonged to King Henry VIII. It appeared to be to have a barrel that was 2 foot 9 inches (0.84 metres) long. A central spindle accommodated a rotating barrel in which there were four bores, each approximately 12.7 mm in diameter.

There was a separate flash-pan for each chamber, covered with a sliding lid, and these were rotated manually in succession under a serpentin (Greener 1910).

Nevertheless, Gatling's basic barrel design lives on today and is still used in systems such as the Phalanx Close in Weapon System. This is a radar guided 20 mm-calibre M61 Vulcan Gatling gun cannon used primarily for defence against anti-ship missiles. It is capable firing 4500 rounds-per-minute. Another example of Gatling's concept is the infamous GAU-8 Avenger which is a 30-mm calibre seven-barrel Gatling system that is installed on the tank-killing A10 Warthog and is capable of firing up to 3900 rounds-per-minute. This rate of fire can only be achieved by a mechanical rotating barrel concept—first invented by Gatling!

Gatling's automatic firing gun was purely mechanical and as such required to be worked by the handle being cranked. However, an American born British inventor, Sir Hiram Stevens Maxim (1840–1916) substantially improved on Gatling's design. Maxim was a genius who never had more than five-years of schooling (Maxim 1962). He was a prolific inventor and was Thomas Edison's chief rival in the development of electrical systems and installed the first electric lights in a New York City building (the Equitable Insurance Company, 120 Broadway) in the late 1870s (Browne 1985). Famously, he narrowly missed out as being acknowledged inventor of the incandescent lamp and in a patent suit Thomas A. Edison proved priority by only a matter of days (Maxim 1962). He also dabbled in flight and may have even beaten the Wright brothers to it if he had used a lighter-weight power source than his steam engine. It was a fateful trip to Europe that changed his fortunes, some would argue for the better. He later wrote to the Times of London:

> In 1882 I was in Vienna, where I met an American whom I had known in the States. He said: 'Hang your chemistry and electricity! If you want to make a pile of money, invent something that will enable these Europeans to cut each other's throats with greater facility. (d 1985).

His compadre was right! The way to make money was in death (and arguably it still is today if you look at the profits of some of the larger Defence contractors). This advice must have also been promulgated to his family as his son (Hiram Percy Maxim) developed a silencer and his younger brother (Hudson Maxim) was connected to the invention of smokeless propellant—allowing guns to be fired with less smoke and thus reducing the risk of alerting the enemy to the shooter's position. Maxim's invention brought death to a mass-produced scale that had never been seen before (see Fig. 2.7). As Lord Salisbury kindly pointed out to Maxim in 1900, he had *"prevented more men of dying from old age than any other man that ever lived"* (Chivers 2011). The Maxim gun also became a key element in colonialization and for the first time it meant that an enemy's number superiority could be overcome by weapon engineering. This was demonstrated just eight years after the invention when Maxim's 'killing machine' was used to slaughter 3000 Zulu Warriors by fifty British security guards of the Rhodesian Charter Company in Africa operating four weapons. Apparently, this all occurred in 90 min which was remarkable for the time given that the temperature of the guns must have been running red hot. As C J Chivers points out, the numbers here are suspicious, but nevertheless the point is made (Chivers 2011).

Such was the fame of Maxim's invention that it promulgated a psychological fear into anyone that was facing a barrage of bullets from its mouth and in 1898 it found its way into Hilaire Belloc's (1870–1953) 'The modern traveller" (Belloc 1898). Belloc's poem was about three Englishmen that travelled to Africa to make money and to exert their will on the natives. One of the characters had the rather fitting name of William Blood and used a Maxim to settle a wage dispute:

> I never shall forget the way
> That Blood upon this awful day
> Preserved us all from death.
> He stood upon a little mound,
> Cast his lethargic eyes around,
> And said beneath his breath:
> Whatever happens we have got
> The Maxim Gun, and they have not.

Maxim's first patents related to the development of the Maxim were registered in June and July 1883. This was the gun that was proliferated in the First World War on all sides and brought death to a whole new level. The German Army's Maschinengewehr 08 as well as the Russian Pulemyot Maxim (PM M1910) were both copies of the Maxim. Simply by keeping the hand on the trigger a spew of bullets was sent forth cutting down men in their tracks.

Maxim was knighted by Queen Victoria in 1901.

Subsequent to the Maxim machine gun came the Vickers machine gun. The Vickers machine gun was an improvement on the Maxim machine gun. Vickers had actually purchased the Maxim company in 1896 and so he had the opportunity to take the design of the Maxim gun and improve on it. This included the use of better-quality steel alloys for certain components; it was fitted to a 0.303 British (7.7 mm) calibre. Like the Maxim, the Vickers was water cooled which made it a

**Fig. 2.7** The maxim machine gun on a tripod and ready to fire (after (Crozier 1916))

solid performer under sustained fire and meant that it was extensively used well into
the 20th C. Water cooling is a significant advantage over modern General-Purpose
Machine Guns as it allows for sustained continuous rates of fire. Modern air-cooled
machine guns have to frequently stop firing in order to change out the barrel—making
the weapon vulnerable during the barrel change. However, it was heavy. The machine
gun typically required between six to eight men to operate it and due to its weight,
was reserved in part for gun emplacements. Yet it had a reliability for being a solid
performer, largely due to the water cooling. For example, in August 1916 it was
reported that there were ten Vickers guns that fired continuously for a duration of
twelve hours. Goldsmith writes an account by Captain Graham Hutchison regarding
an attack on High Wood:

> It is amusing today to note that in the orders for the 100th Machine Gun Company's barrage
> of 10 guns, Captain Hutchison ordered that rapid fire should be maintained continuously for
> twelve hours, to cover the attack and consolidation. It is to the credit of the gunners and the
> Vickers gun itself that this was done! During the attack on the 24th, 250 rounds short of one
> million were fired by ten guns; at least four petrol tins of water besides all the water bottles of
> the Company and urine tins form the neighbourhood were emptied into the guns for cooling
> purposes; and a continuous party was employed carrying ammunition. (Goldsmith 1994)

In 1889 John Moses Browning (1855–1926) was also working on a new methodology for automatic firing. However, Browning was pipped to the post by Maxim.
Nevertheless, Browning's focus was on hand-held weapon systems with his most
successful design being the .45 calibre M1911 single-action semi-automatic recoil-
operated pistol. This entered service with the United States Armed Forces in 1913
and was phased out by the M9 in the late 1990s. In fact it is said that there are special
operation variants still in service today and will be for some time (Rottman 2013).
Browning had noticed that a rush of gas followed the bullet and that this rush of
gas was powerful enough to flatten the bulrushes in the marshes in Utah—his home
State. And so Browning started to work to harness some of that gas and use it for
kinetic effect. The end result was his patent of the M1895 Colt Browning machine
gun.

Again, the basic principle is quite simple: namely that a hole some way down the
barrel allows to vent some of the excess propellant gases which then in turn act on
a pusher rod. The pusher rod, in turn brings the bolt backward and at the same time
expelling the spent cartridge to allow for the feed of a new round.

Harnessing the kinetic energy of the hot propellant gasses is by and large what is
done today with the more modern assault rifles and semi-automatic shotguns. This
led to the BAR or the Browning Automatic Rifle. This saw extensive use towards the
tail end of the First World War right through to the Vietnam conflict in the 1960s. It
was highly regarded as a weapon. It was highly reliable; it contained a 20-clip box
type magazine and a shooter would have no problem walking forward and attacking
the target with it. Suffice it to say it was a very popular weapon for US Marines.

The key to a good understanding of how to design a gun for automatic operation
comes from a knowledge of the pressure time curve for a propellant. As the propellant
deflagrates (which, is another word for 'burns') the pressure builds up rapidly in the
chamber as hot propellant gasses are formed—such as carbon monoxide, hydrogen

and nitrogen. These gases start to act on the projectile and begin to push the projectile down the barrel. For rifled barrels, the projectile embeds into the rifling by virtue of its jacket being scored by the strong rifling present in the barrel. And so, as the force required to squeeze the bullet into a grooved bore are overcome, the projectile starts to accelerate. Accelerations of up to 100,000 g are quite possible. As the projectile moves, the volume expands and thereby accommodating a drop in the pressure behind the bullet with it. Eventually the propellant is completely consumed by the burning process and the gas continues to expand and as it does so the projectile continues to accelerate. By the time the projectile passes the vent holes for recycling the weapon the local pressure has dropped off considerably.

The positions of the vent holes are also important. Too close to the cartridge and the local pressure will be too high thereby forcing piston back with too large a force so as to cause material failure. Too far from the cartridge and there isn't enough pressure to recycle the weapon. The same could be said for their size: too large and the recycling force becomes too high; too small and the gun is unable to recycle in a timely fashion.

## 2.6  Automatic Versus Semi-automatic

The use of gas recycling had led to a whole new wave of products, some of which were automatic, such as Maxim's machine gun and semi-automatic weapons. Some confuse semi-automatic weapons with fully automatic weapons and it is important to remember that what is automatic is the ejection of the spent cartridge and replenishment of the next round ready for firing. So, a semi-automatic weapon fires a single shot for every pull of the trigger—however, of course that can be rapid. An example of a classic semi-automatic weapon was the SKS assault rifle that was developed for the Soviet Union during World War II. This semi-automatic weapon was still capable of firing 35–40 rounds per minute with multiple pulls of the trigger. Controversially, these types of semi-automatic weapons can be converted into fully automatic weapons using what is called a 'bump stock'. This being a device that is used to replace the existing stock of a semi-automatic weapon and literally 'bumps' the whole gun forward such that the trigger is pushed forward onto the finger of the shooter.

## 2.7  Cartridge Evolution

An important part of the way that guns work revolves around how the propellant is contained in the chamber. Early cartridges were made from paper and ironically modern tank guns still use a combustible case similar to paper. Of course, all small arms and many recent larger guns have used brass cartridges to house the propellant. Brass is an alloy of copper and zinc in the ratio of 70% copper and 30%

zinc. It is an alloy that is inherently corrosion resistant although given the wrong storage conditions can result in a premature burst in a gun chamber. This is due to a process called stress corrosion cracking where internal residual stresses caused by the manufacturing process provide a condition for cracking to ensue, given the right corrodent. This was especially true during the days of the Raj where gun cartridges were stored near horse stables. During a rainstorm the urea from the horse would act as a corrodent and actively penetrate the cartridge brass thereby compromising its strength. Nevertheless, the solution to this problem was simple and low temperature annealing was introduced which involved heating the brass up to 250° centigrade and being allowed to cool slowly. This processing relieved the internal stresses of the brass and therefore stopped the process of stress corrosion cracking. Nowadays there is not much concern for this—mainly because we do not tend to store huge hordes of ammunition close to horses. Nevertheless, sodium chloride (common salt) is another corrodent and therefore naval ammunition requires a final production heat treatment to avoid this type of failure, often indicated by the initials LTA stamped on the base of the cartridge (Doig 2002).

The material science of the cartridge case is really important and early cartridge designs were prone to failure and caused jamming in early Gatling and Maxim models of rapid-fire weapons. The choice of the alloy is particularly important as there is a need for an alloy that can accommodate the stress on firing during ignition and at the same time it can relax once the firing process has ceased to allow for extraction. Small clearances are required to allow for insertion into the chamber (and for a 105 mm gun this would be around 0.7 mm). Brass was commonly used as it is a material that can be easily worked and forged into the correct shape through a series of processes. Brass has approximately half the elastic stiffness of the steel gun barrel and a high strength is crucial for maintaining performance and ensuring the elastic recovery does not result in jamming.

It is also important that sufficient sealing of the propellant gasses occurs where the cartridge meets the projectile so that when the propellant ignites, the full force is transferred to the projectile. At the same time, it is also important that the rear of the cartridge is strong so that it can be gripped and pulled out of the chamber by the extractor. So, for many larger calibre cartridges there had to be a hardness gradient along the length of the cartridge to accommodate this principle requirement. That is, the base was hard whilst the neck was malleable.

Brass cartridges are generally quite expensive and there has been a concerted effort to try and reduce the cost of manufacture. The expense mainly come from the brass material which comprises of two very expensive elements (Cu and Zn). To replace brass, you would need a material that would have the same elastic properties as well as similar ductility and thermal properties. Brass is a good conductor of heat thereby allowing the heat to dissipate out into the steel chamber. Of course, for larger guns there has been a tendency to move away from brass cartridges altogether. The L7 105 mm tank gun that was fitted to the Leopard 1 Main Battle Tank was one of the last modern-era tank guns to use a brass cartridge and most modern tank guns use combustible cartridge cases. Obturation (that is, sealing) is provided by a stub case. An additional disadvantage was that the Leopard 1 had a little port in the turret that allowed for the gunner to dispose of the cartridge after firing. Of course, the crew wanted to avoid being dwarfed in a mangle of hot brass cartridges in the turret. This was not particularly helpful given that the opening of the port compromised protection against nuclear, biological and chemical attack. In addition, due to the cost of the cartridges there would be an expectation that the crew would pick up the cartridges after their training exercise so that the cartridges could be recycled. Nice! Low carbon steel cases are sometimes used and other materials have been put forward such Cu-Be, which would cost six times that of cartridge brass but provide twice the strength, making it a possible candidate for high-performance choices. More recently there has been a trend to examine the viability of high entropy alloys to replace the expensive cartridge case materials. These are alloys comprising of numerous alloying elements, each roughly in a similar wt% composition.

The cost of cartridge brass has led to a common pastime among shooters in reloading ammunition from previously spent cartridge cases. Recycling of cartridges is a whole other industry and manufacturers of cartridges cater for the those who wish to undertake self-loading of their ammunition. Small arms cartridges can be re-loaded several times (up to 10 times, in some instances). The process involves fitting a separately supplied primer cap, propellant and projectile to the cartridge and usually involves a process of trial and error to establish the right 'fit' into the chamber. It is quite an art-form! The cartridge will tend to stretch with each reload and consequently the wall will tend to thin. So, the length of the cartridge will need to be trimmed from time-to-time to ensure that the cartridge sits correctly in the chamber. If this is done incorrectly then this will give rise to headspace issues. Headspace is the distance as measured from the front of the bolt face and the part in the chamber that stops the cartridge moving forward. Too much headspace and the there is too much space in the chamber that could allow for the cartridge to stretch in the direction toward the bolt face, thereby risking case-head separation. This is a serious event and can lead to injury of the shooter. Too little headspace and the bolt may not close. So, getting the headspace measurement correct is important and that is why the cartridge case needs to be carefully trimmed. In any event, weapons come equipped with vent holes that should adequately vent the high pressure in the circumstance where failure of the cartridge has occurred.

Obturation is profoundly important. If you cannot seal the chamber to stop propellant gases escaping, then people's lives are at risk. A failure to seal the propellant gases has hurt, maimed, killed many a shooter through history. Bear in mind that pressures in the chamber of a sniper rifle propellant gases can reach 4000 bar, in a tank gun, it is more like 6000 bar and this can spew out rapidly high-temperature gases.

And that sadly, is what happened to Experienced Army trainers Corporals Matthew Hatfield and Darren Neilson on 14 June 2017. These gentlemen died of their injuries after the gun in the Challenger 2 they we operating spewed hot propellant gasses into the turret. It turned out that the L30A1 120 mm rifled gun they were operating did not have a key piece of equipment called a BVA assembly (Bolt Vent Axial). This is a mushroomed-shaped assembly that plays a key role in sealing the propellant gasses once the gun is fired. It fits into the breechblock, which for the Challenger

2 is a unique two-part construction. As the charge burns, the pressure builds on the front face of the BVA, forcing it rearward and compressing an obturator pad, which is a made from a compliant heat resistant material, against the breech block. This forces the obturator pad radially outwards to create a gas-proof seal between the obturator pad and the rear of the barrel chamber. All this would have happened very quickly and usually a firing is over in a millisecond. The two-part breech-block was compromised and became separated with the top part finding its way to be wedged in the roof. The soldiers' fate was further compounded by the fact that it seemed common practice for armourer and tank commanders to store propellant charge bags on their laps so as to maintain a high tempo in firing (Felton 2018; Hunt 2018).

The use of stub cases in main battle tank guns has now become the norm rather than the exception. And it is the stub case that is all important. Because the length of the stub case is small compared to the length of the chamber, it is not unreasonable to use steel for the stub case. This also has the added advantage that in a confined turret only a small object must be disposed of.

## 2.8  Bolts

For rifles, obturation is accomplished by the cartridge case being kept in place by a bolt. The majority of bolt-action rifles employ a turn-bolt approach whereby the shooter will execute a forward motion to push the ammunition into the chamber followed by a downward action to lock the bolt in place. Locking is usually accomplished by the rotation of robust locking lugs into notches in the receiver body. This is necessary so that when the projectile is fired, the cartridge case assembly (which includes the cartridge case and primer cap) is retained securely in place.

An alternative form of locking mechanism is employed in a straight-pull bolt system where there is no need to rotate the bolt handle to secure the bolt in place. This allows for increased speed of operation compared to conventional bolt-action rifles. As the ammunition is fed from underneath the weapon, the action of pushing the handle forward (attached to the bolt head assembly) allows for picking up the ammunition and pushing it into the chamber. To lock the bolt in place, one final push on the handle is required and this simultaneously opens some form of locking mechanism. Any safety mechanism that is in place to stop the gun firing is disengaged and the firing pin would then be allowed to strike the primer cap once the trigger is pulled. This type of 'action', as it is called, is favoured by military forces mainly because of the speed of operation. For example, the Blaser (or now XTEC) TAC 2 in service with the Australian Defence Force uses such a system.

The bolt assembly, that includes the ejector post and the extractor can see immense forces during the firing cycle and therefore need be resilient and so consequently they are made from hard and tough steel. If the bolt fails then serious injury will occur. Probably one of the simplest yet ingenious mechanisms on a gun is the extractor. This is the device that locks onto the cartridge once the ammunition is loaded into the chamber by the bolt. All small-calibre cartridges used in a hand-guns and the like have a rimmed base to which a 'springed' claw located on the bolt head grips onto it once the bolt is driven forward. Once the bullet has been fired, the bolt is retracted and the cartridge is 'flicked' out of the weapon through various mechanisms. Commonly this process is via an ejector post that is sat in the bolt head.

## 2.9   Breeches

For larger guns it is common that a breech is used to allow access to the bore and provide a resilient locking mechanism to cope with the rear-acting forces during projectile acceleration. A breech is essentially the 'window' to the bore. A breech assembly will comprise of two major items: the breech ring—which is a structural item that is directly attached the barrel and, the breech block that can either be in the form of a sliding breech block or a breech screw.

The first breech screw system with an interrupted thread was developed by Charles Ragon de Bange (1833–1914) who was a French artillery officer (see Fig. 2.8). Up to the point of de Bange's invention, breech-loading cannons had not been providing reliable obturation to the hot propellant gasses that would form on ignition. Thus, muzzle-loaders were still favoured. Prior to this, there were some attempts made by Martin von Wahrendorff (1789–1861) who developed a breech-loader by using a cylindrical steel plug, secured by a horizontally applied wedge. Equally, William Armstrong (1810–1900) developed a similar system whereby a laterally applied sliding block was pressed up against the face of the bore with a screw breech. None of these systems were satisfactory.

De Bange's development, however, was satisfactory and could be operated in reasonably fast manner and it meant sizeable breeches could be sealed effectively with a breech screw attached to what best can be described as a door that would swing open to receive the ammunition. The main advantage of the design was the addition of a doughnut-shaped grease-impregnated asbestos pad that sealed the breech. Asbestos

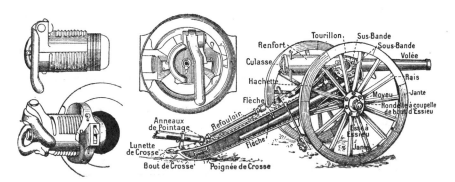

**Fig. 2.8**  De Bange's breech loader cannon concept with the interrupted thread. The cannon had a 90 mm calibre (ca. 1877)

was the new 'wonder material' of the time due to its superior heat resistance. The asbestos pad was equipped with a separate rounded nose cone. When the gun fired, the nose was driven rearward into the asbestos pad, thereby compressing it. This led to the breech's sealing. These, however were somewhat temperamental in performance and little reliance could be placed on prolonged use of the obturator due to its limited resilience. So, maintenance was key. It was standard practice for pads to be removed from the guns, during transit and periods of inaction.

The design of the breech was further refined by Ernst Martin Axel Welin (1862–1951) with a stepped interrupted thread with a progressively increasing radius, that provided more thread onto which the loads could be applied. The principal advantage was that the breech screw element could be shorter. The Welin breech was a single motion screw, allowing it to be operated much faster than previous interrupted-thread breeches. It soon became very common on British and American large calibre naval artillery (see Fig. 2.9).

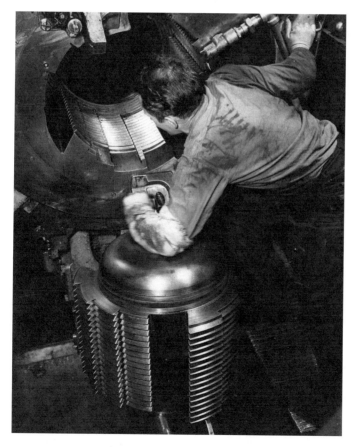

**Fig. 2.9** A sailor looking into the breech of a 16-inch gun on the USS Alabama. Note the Welin breech screw with its ever-increasing radii of threads (Public Domain)

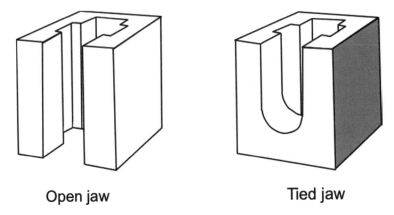

**Fig. 2.10** Two types of breech rings that would receive the sliding breech block. *Source* Author

However, in confined spaces, the breech screw was problematic as the swept volume of the breech door can interfere with activities. And so, in turrets a sliding assembly is usually preferred. This is to limit the volume occupied by the opening of the breech to receive the ammunition.

For the sliding assemblies the breech ring is commonly either of the form of an open jaw or tied jaw (see Fig. 2.10). The open jaw is cheaper to make as there is a clear path for a cutting tool to move through the steel. On the other hand, the tied jaw example is more expensive and requires complex wire erosion techniques to cut out the desired shape.

Unlike the breech-screw approach, the sliding block system is reasonably difficult to obturate. This is due to the fact that to allow the sliding of the steel block within the breech ring there has to be clearance that precludes the use of a positive seal. Therefore, most sliding systems have been used with cartridges or stub cases that expand during the deflagration phase. However, for the L11 Challenger 1 gun, there was a requirement to maintain obturation by the breech design alone and this was achieved by using two carefully machined slip rings: one in the breech block and one in the sliding breech block. The rings themselves had to have a less-than 10 micron finish to ensure correct functioning and were lapped (ground) using a specific type of diamond paste. When these were brought together, a positive seal would be achieved. The trouble with this however was that in the event that the cotton propellant bags frayed, then the cotton could stop the breech from closing completely. Even small gaps would be disastrous. To get over this problem, the British engineers devised a solution whereby a metal safety pin would protrude to prevent the ammunition being loaded in the event that the metal sealing rings were not in place or correctly seated.

Another solution was to use a split breech block. This was eventually used in the Challenger II and was a unique take on the sliding breech block. It made use of both the advantage of a breech screw arrangement as well as the speed and the benefit of the reduced volume of the sliding block approach. Effectively a set of lever mechanisms were able to pull one half of the block away from the bore thereby facilitating the addition of a mushroom head obturator fondly referred to as the 'Crossley pad'. The Crossley pad, named after its inventor, Arthur Crossley who invented it in 1944, comprised of a neoprene construction that when compressed by the expanding propellant gases provide satisfactory sealing (Crossley 1944). As the breech block was pulled rearward, the rear part of the block dropped downward to facilitate the rear movement and thereby allow for the opening of the breech.

More recently, a different take on breech design has been introduced. The CTA (cased telescopic ammunition) is an approach whereby the ammunition is fed perpendicular to the barrel axis and rotated through ninety degrees to fire. The whole basis of this breech design is to replace the conventional ammunition design by containing the projectile, sabot (if required), propellant and all of the cartridge ancillary components in a single cylinder. This type of approach is probably the future of ammunition.

# References

Belloc H (1898) The modern traveller. Basil Temple Blackwood, London

Browne MW (1985) 100 years of maxim's 'killing machine'. The New York Times, 26 November 1985

Chivers CJ (2011) The Gun. Simon and Schuster, New York, USA

Crossley A (1944) Obturator for ordnance. US Patent

Crozier W (1916) Handbook of the maxim automatic machine gun calibre. 30, model of 1904 with Pack outfits and accessories. War Department, Office of the Chief of Ordnance, Washington

Doig A (2002) Military metallurgy. Maney Publishing, London, UK

Felton RFP (2018) Service inquiry—Challenger 2 incident at Castlemartin ranges, Pembrokeshire. Defence Safety Authority

Ffoulkes C (1969) The gun-founders of England: with a list of English and continental gun-founders from the XIV to the XIX centuries. Arms and Armour Press, London, UK

Goldsmith DL (1994) The grand old lady of no man's land: the Vickers Machinegun. Collector Grade Publications, Cobourg, Canada

Greener WW (1910) The gun and its development, 9th edn. Bonanza Books, New York, USA

Hunt L (2018) Senior coroner's report—regulation 28 report to prevent future deaths. HM Senior Coroner's Office, Birmingham and Solihull, Birmingham

Kelly J (2004) Gunpowder: alchemy, bombards, and pyrotechnic. Basic Books, New York, USA

Maxim HP (1962) A genius in the family. Dover Publications Inc, New York

Olcer NY, Lévin S (1976) Recoilless rifle weapon systems. U.S. Department of Defense, Army Materiel Command

Rottman GL (2013) The book of gun trivia: essential firepower facts. Osprey Publishing, Oxford, UK

# Chapter 3
# Fast Guns

As I mentioned earlier, high-velocity rifles can fire projectiles to over 2000 miles-per hour. However, this is not the fastest that guns can fire projectiles. In this chapter we will review some of the faster guns, even guns that can launch projectiles into space! Science fiction? Well, let us see what the science says…

## 3.1 Why Faster?

Currently tank guns are able to fire projectiles to reasonably high velocities that can result in kinetic energies of up to 10 megajoules. To put that into context, this is similar in energy to firing five Toyota Landcruisers (from a very large gun) to a speed of 100 mph. And, all of that energy is concentrated over the target in an area that is similar to that of a dollar coin. So why shoot faster?

Arguably the need to go faster is driven by two aspects. The first of these is to do with 'lethality' or our ability to kill[1] a target. During the Cold War there was a push on tank-gun development that led to some prototypes of 140 mm tank guns, which was an increase in size on the 120-mm guns. There were also several programs to develop rail-gun technologies, that seemed to ebb and flow. The main objective was to defeat the perceived increase in protection that would have been offered by the then Soviet armour, simply by accelerating a longer rod to a higher velocity. Once the Cold War was over, most of the 140-mm gun programmes were cancelled. The second aspect to the requirement for faster projectiles is range. High-velocity projectiles, in particular supersonic projectiles, are subjected to very high drag forces during flight. One way that the range can be extended is by assisting the projectile with a rocket attached to it. However, this is very difficult to do with gun-fired projectiles.

---

[1] Although this sounds rather morbid, the military frequently focus on ideas such a 'mobility kill' where the objective is to simply stop the enemy moving. This would be included in what I mean by the word 'kill'.

© Springer Nature Switzerland AG 2021

P. J. Hazell, *The Story of the Gun*, Springer Praxis Books,

https://doi.org/10.1007/978-3-030-73652-1_3

One of the most successful ways of increasing the range of a projectile is simply by increasing its muzzle velocity from the gun. For example, the main purpose of the US Navy's rail-gun program was so that they could put "steel on target" at ranges of more than 100 km. This would enable their warships to provide support to troops on the ground.

## 3.2  Limits with Conventional Guns

Propellant-based guns have a natural limit in that they are unable to maintain the push behind the projectile for the length of the gun barrel. This is because of the way propellants burn. They generally burn rapidly (over a few milliseconds) and as they do they produce hot products of combustion that expand and therefore push the projectile. However, the push is short-lived. As the "push" is not maintained and cannot be sustained for any great distances, the projectile stops accelerating.

So, what happens to the propellant during its violent but short life? During the burn process the chemical energy is released into the kinetic energy of the combustion products in the formation of fast-moving molecules. This is achieved by burning nitrocellulose-based compounds that result in the production of relatively heavy carbon and nitrogen-based molecules. These molecules act against the base of the projectile to give it the necessary push. So, can we achieve high velocities with powder guns?[2] Certainly, with a double-length barrel and a large amount of propellant, powder guns are able to fire light-weight projectiles to close to 2.8 km/s. To achieve higher velocities the answer does not lie in packing more propellant into the cartridge as there is a limited return on investment for increasing the quantity of the propellant in the breech. The reason for this is that the internal energy of the propellant is divided between the kinetic energy of the gas molecules and the kinetic energy of the projectile. The more propellant you add to the chamber, the higher the mass of the gas molecules that are accelerated. Eventually you find that the mass of the propellant gasses formed exceeds that of the projectile. Therefore, there is a law of diminishing returns at play! Furthermore, propellants are seemingly inefficient because their chemical energy is never solely passed onto the kinetic energy of the projectile or even the propellant gasses. In fact, much of it is wasted. Much of it is turned into heat, for example, and lost to the barrel (as those who fire machine guns will know all too well!).

There are various thoughts on how you can improve the performance of conventional guns and that starts with dealing with the energy density of the propellant. One possible solution would be to find a propellant that produces low molecular (lightweight) driving gasses as the propellant burns. Thus, the internal energy of the propellant is not all consumed in accelerating the gas molecules. However sadly, nature has not been kind enough to reveal an obvious solution for that yet!

---

[2] A 'powder gun' is a reference to the guns that once used gunpowder as the main propellant source. Of course, as we have seen this is no longer the case however the term has stuck.

The energy density, that is the amount of energy that is locked away in 1 kg of propellant is around 1 MJ/kg. As it happens petrol has an energy density of about 45 MJ/kg, so why don't we use gasoline to propel projectiles? Well, in some ways we do—in the internal combustion engine. Except in this example, the projectile (the piston) is connected to a crank shaft via a connecting rod, as already discussed. Furthermore, gasoline needs a substantial amount of oxygen to create the reaction that drives the engine (around 15 kg of air is required for every kg of gasoline consumed). Hence that is why your local mechanic recommends that you change your air-filter periodically! Propellant has its own oxygen and so the energy densities are about on par when you consider the air that is required for gasoline combustion. Finally, you need to release that energy quickly and safely within a pressure vessel (cartridge) and that is where nitrocellulose-based propellants are appropriate and have an advantage over gasoline.

Tank guns boast the fastest velocity of all military guns and for good reason. They need to fire projectiles with sufficiently high kinetic energies to penetrate the enemy's armour. Furthermore, they frequently attack moving targets and therefore once fired, the projectile can hardly hang around before getting to the target. Tank gun velocities generally top out around 1900 m/s (Mach 5.6). To achieve this with heavy projectiles means that the propellent is often supplemented with RDX (a high explosive). This severely limits barrel life and it is not unusual for barrels to only last 200–300 firings at maximum charge[3] before they need replacing.

## 3.3 Lessons from World War II

The desire for higher velocities was no more pertinent than during World War II. The Allies were somewhat stumped when it came to defeating the enemy armour—especially when it came to the mighty King Tiger tank. The Tiger employed rolled homogeneous nickel-steel plate armour which was strong and tough. Furthermore, the German Tiger II of 1944 had a 150-mm plate at the front of its hull and a 185 mm plate at the front of its turret. A similar vehicle, the Jagdtiger which was based on the Tiger II chassis and had a 128-mm gun, had an even thicker plate of 250 mm of armour on the front section of the tank. No single homogenous armour plates of that thickness have since been employed on tanks and the Jadtiger was the heaviest armoured-fighting vehicle of World War II, weighing in at around 71.7 tonnes (Ogorkiewicz 1991b). Hence, these tanks were difficult to defeat with conventional weapons available to the Allies.

There were two options available to the Allies for defeating this heavy armour. One was to develop a sabot system that allowed the pressure from a large bore cartridge to act on the base of a sabot that contained a sub-projectile. The sabot would carry the smaller-diameter sub-projectile out of the gun. This became known as the Armour-Piercing Discarding Sabot (APDS) projectile. The second option was

---

[3] 'Maximum charge' sometimes referred to as 'effective full charge' or 'EFC'.

to use what became known as the squeeze-bore shot using a so-called, "Littlejohn[4] adaptor" fitted to existing gun barrels. Here, deformable flanges surrounding the projectile were squeezed by the adaptor during launch. This allowed for a relatively large area for the propellant gases to act on initially in the chamber, whilst allowing for reduced diameter and hence better penetrability at the target. The theory was sound. However, these had limited success as the adaptor was subjected to high frictional loads and considerable energy derived from the propellant was required to squeeze the projectile down to size (see Chap. 7). Consequently, the sabot system was perceived to be the best option.

During the years of 1941–1944, the APDS technique was developed in the United Kingdom by L. Permutter and S. W. Coppock working at the Armaments Research Department. They developed a system comprising of a low diameter tungsten-carbide core that was carried along the gun barrel by aluminium 'collars' that were centrifugally discarded when they left the gun barrel. The core would then simply fly on to its target without much velocity loss. In mid-1944 the APDS projectile was first introduced into service for the UK's 6-pounder[5] 57 mm anti-tank gun, as employed on the Churchill tank, and extensively used in Normandy. The 17-pounder saboted projectile was introduced later in September 1944. The velocity of the projectile fired from a 17-pounder tank gun was around 1,200 metres per second. However, the muzzle velocity of APDS ammunition produced for the 83.8-mm 20-pounder which was introduced in 1948 had increased to 1465 m/s. And, in the 1950s an APDS projectile developed for the 105 mm L7 gun had a similar muzzle velocity of 1478 m/s. This represented the highest velocity of APDS ammunition in regular use. That is, except for the version of the L7 gun that was employed on the Swedish S-tank which had a slightly longer barrel and therefore was able to achieve just a notch over 1500 m/s (Ogorkiewicz 1991a). We will discuss this again in Chap. 7.

## 3.4  From Single-Stage to Two-Stage

All commercial military guns are what we call 'single-stage guns'. That is to say that they employ one breach and one barrel. However, in the 1940s it was realised that it was possible to build a gun that used two breaches and two barrels. The first concept was invented by William Crozier during World War II and subsequently patented in 1959 (Crozier 1959). Crozier was working under the guidance of a Professor Workman at the New Mexico School of Mines at the time. The basis of this invention was to use lightweight gasses as the driver gas to accelerate the projectile. To do this Crozier took a reservoir of lightweight gas (e.g., hydrogen or helium) and dynamically compressed the gas using a propellant-launched piston. In the second stage (hence 'two stage') the projectile sat and kept in place by some form of diaphragm or

---

[4]"Littlejohn" was the Anglicisation for the name of the Czech inventor.

[5]The tradition of classing guns by the mass of the projectile they fired dates back to an edict issued by Henry II in 1550.

breakable flange. As the pressure from the low molecular gas continued to increase, eventually the flange broke and allowed the projectile to transit down the gun barrel at high velocity.

High-velocity guns were particularly important during the 1950s—mainly because there was a lot of interest on how ICBMs re-entered the atmosphere. These missiles, which could travel up to 16,000 km would fly into space and re-enter the Earth's atmosphere at between 6000-8000 m/s. During re-entry the missile body becomes hot due to aerodynamic heating and in the 1950s, experimental facilities were lacking that would allow scientists to study the effect of aerodynamic heating. Early work at what became NASA's Ames Research Center, focussed on attempting to get a gun to fire projectiles beyond 4 km/s.

The main challenge was to achieve high velocities whilst still ensuring that the relatively fragile models of re-entry vehicles remained intact during flight. The ideal pressure profile behind the projectile was thought to be a smooth ramp up to the maximum allowable pressure followed by a constant level and then a smooth ramp down near the muzzle. Essentially, the pressure time response in the barrel should look something like an isosceles trapezoid. This was best achieved by the integration of a conical section at the end of the first stage or 'pump tube' (Charters 1995).

Amazingly these kinds of two-stage guns (or modifications of them) have been known to fire projectiles to 8 km/s. So how do they work? A layout is shown in Fig. 3.1. The principle of the gun is that a piston is fired in the first stage of the gun which holds a gas held under a small pressure—the size of pressure that you might want to pump into your bicycle tyre, for instance. The movement of the piston, which itself is accelerated by a propellant charge or by a the rapid release of a high-pressure reservoir of gas, compresses the lightweight gas and accelerates it through a conical channel where the pressure reaches an enormous pressure very quickly. This pressure is magnitudes higher than bike-tyre pressure. That is, unless your bike tyres are made from high-strength alloy steel! Behind this secondary breech there sits a projectile, usually saboted, and sitting snuggly in a smooth bore gun. The projectile is accelerated very rapidly and can experience huge g-forces as it finds its way to the target.

Apart from testing the flight of fast-moving missiles, there was another motive for developing these high-velocity guns. Namely, missile defence. If you could launch a spray of fragments to a high velocity such they perforated the re-entry shield of a missile then you could render the missile useless. The problem was at that time the effect of such hypervelocity penetrators on the target was unknown. Moreover, the era of space-flight was just beginning and the safety of space vehicles and their occupants was of huge concern. Not only do they have to contend with the vacuous nature of space, there are huge temperature variations and radiation to contend with. In addition, an acute challenge for orbiting space craft was the millimetre-sized meteoroids that could smash into space vehicles at velocities of up to 20 km/s. Despite their small size, these tiny meteoroids would have a devastating effect on a space vehicle. This has more recently been compounded by the almost exponential

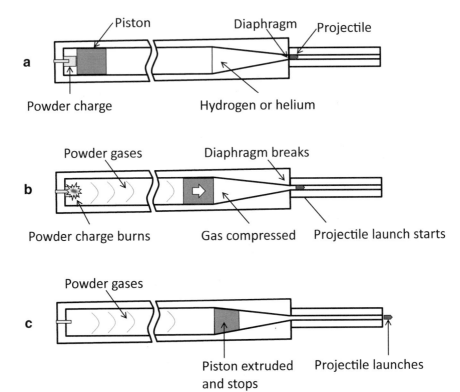

**Fig. 3.1** Two stage gas-gun operation. **a** The pump tube is filled with a low molecular weight gas. **b** This gas is compressed by a propellant charge deflagrating and producing a pressure that accelerates the piston towards the conical section. **c** The piston is extruded into the conical section and stops; the projectile is accelerated by the light gas. Adapted from (Charters 1995)

growth in space junk, as fans of the Sandra Bullock/George Clooney film 'Gravity'[6] will know.

However, Fred Whipple (1906–2004), an astronomer at Harvard, had the solution. Whipple postulated that by using a thin layer of metal spaced a short distance from another layer it is possible to sufficiently armour the space craft against these fast-moving projectiles (Whipple 1947). It turns out that the high velocity of the meteorite is its own undoing. It can be smashed and dispersed by a relatively thin layer of metal about the thickness of meteorite itself. The outer two layers of thin metal sheet became known as the 'Whipple bumper'. The two-stage gun helped optimise Whipple's designs and provided the test apparatus for simulating fast-moving space projectiles.

Figure 3.2 shows how the gas guns evolved over the years with the maximum velocities shown by the black circles. These represented the maximum velocities

---

[6]Warner Bros. Pictures, directed by Alfonso Cuarón, 2013.

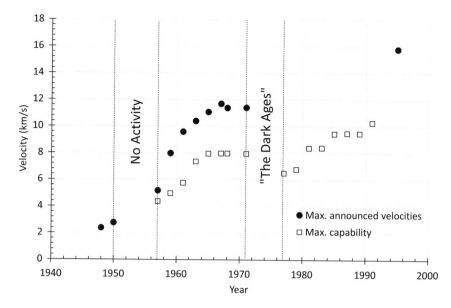

**Fig. 3.2** Reported peak light-gas gun velocities versus time from 1948. Adapted from (Swift 2005) with added data

attained by labs which in themselves were frequently not very useful projectiles in that they were small polymer cylinders or plates. And so, the maximum capability 'claimed' by laboratories (to do something useful) was somewhat more modest. Probably there is one exception to that and that is the final data point close to 16 km/s that was an aluminium flyer-plate launched in a three-stage configuration. We will talk about that now.

## 3.5 And Even to Three Stages

It is entirely possible to increase the velocity of a gun beyond 8 km/s by adding a third stage. This work was pioneered by Lalit Chhabildas at Sandia National Laboratories (Chhabildas et al. 1993; Chhabildas et al. 1995). In simple terms, the barrel of the two-stage gun is extended with an expendable section made from a high-density material (tungsten) in which sits a stationary projectile located in a 'guard ring'. A heavy projectile is accelerated towards the stationary projectile where it collides with the relatively lightweight projectile and through a process of momentum transfer, is able to accelerate to an enormous velocity. Velocities approaching 16 km/s are quite achievable (Chhabildas et al. 1995).

The concept here is very similar to a game of billiards where a stationary ball is accelerated by a cue ball. There were clearly several technical problems associated with this, not least the fact that the launch projectile would be shocked and therefore damaged during the collision. The violence of the contact would have been quite staggering! It was reported that the energy required to launch the projectile to 16 km/s was approximately 10–15 times the energy required to melt and vaporize the projectile. Therefore, the energy must be transferred in a well-controlled manner to prevent melt or vaporization. This was largely overcome by using a graded impactor that provided for a relatively soft launch as well as a buffer that acted in a similar fashion to a shock absorber. Thus, the projectile was not shocked per se but given a controlled push—albeit extremely quickly. Despite these technical challenges and the expense (replaceable tungsten barrels are not cheap), it provided a useful tool for space debris impact studies and the like.

## 3.6   Explosive Guns

OK, so let us up the ante. With a slight modification to the two-stage gun, it may be possible to achieve velocities close to 30 km/s (Glenn 1990). This can be achieved by using an explosively driven pusher-plate to compress a lightweight gas, which in turn, drives the projectile. The basic concept uses a 'Voitenko compressor' or 'implosion gun' technique which produce fast moving shock waves in a gaseous medium (e.g., hydrogen). A schematic of the concept is shown below in Fig. 3.3.

Like the two-stage gas guns, the explosive composition detonates and drives the metal pusher plate forward thereby compressing the low molecular weight gas. As it does so, shock waves are formed in the gas which in turn are used to accelerate the projectile, which is initially held in place by a flange. Similar to light-gas guns, these

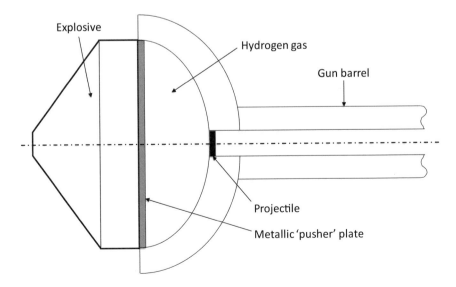

**Fig. 3.3**  A schematic of Voitenko implosion gun  Adapted from (Glenn 1990)

would not be used as weapon systems, but used as part of an experimental apparatus to test hypervelocity impacts. An intriguing application of the Voietenko gun is that it could be used to initiate nuclear fusion.

More recently there has been some work on what is called an implosion-driven hypervelocity launcher by researchers at McGill University in Québec, Canada. These are guns that use explosives to shock compress a light gas such as helium so that it subsequently drives projectiles to velocities exceeding 10 km/s$^2$. Experiments to date have worked with masses of projectiles that range between one and fifteen grams and the researchers believe that there is very good scalability (Hildebrand et al. 2017). This means that it is possible to increase the mass of the projectile whilst maintaining the high velocity. At the time of writing, work is currently underway to increase the velocity of the system. They work is based upon some developments by Physics International in the 1960s (Watson 1970) and uses the implosion of a thin wall tube of metal to dynamically compress a low molecular weight gas, such as helium. As a result of this, the helium is subjected to enormous compression such that a shockwave is formed that goes on to set the projectile in motion. The subsequent compression and then expansion of the helium gas, which is one of the gases that is commonly used, pushes the projectile to very high velocities. As of 2017 the program at McGill University has achieved milestones of 10.4 kilometres per second for a 0.3 g projectile and 7.6 kilometres-per-second for a 50 g projectile. However, one of the drawbacks of this particular system is that it is a one-shot device. And so, it is best suited for experimental campaigns. Nevertheless it is claimed that the cost of replacing the tube and the associated barrel components for each subsequent firings is relatively small (Hildebrand et al. 2017). Still, not very useful in combat!

## 3.7 Electric Guns

There has been a substantial shift in recent times to develop electrically-based vehicles for warfighting. Electric vehicles are becoming more common-place on our roads and even though all-electric armoured vehicles may seem like a pipe dream there are very good reasons to develop such systems. The main reason is due to the logistical burden of fuel. Fuel places a large burden on battle forces and therefore the less gasoline that has to be transported to the battle-zone the better. So, there has been developments in all-electric drives and with that has come developments in electric armour and of course, the electric gun.

Electric guns provide the opportunity for very fast velocities as they are not hindered by the limitations of chemical propellants, although there are other challenges as we will see. Electric guns (rail guns, coil guns and the like) have been around since the time of the development of the electric linear motor and over the past century there has been several notable developments. Even in my home town of Canberra, Australia a rail gun was being developed in the 1970s at the Australian National University (Barber 1972). The invention of the modern rail gun is credited to the French inventor Louis Octave Fauchon-Villeplee. He invented an electric cannon in 1918, based on the principles of the linear motor. The was also a design for an all-electric anti-aircraft cannon that was being developed by the Nazis during World War II although this was never fully realised, by all accounts.

The Office of Naval Research has been behind most recent developments that also included contributions from international partners as well as industry. The benefit of applying this technology to a Naval platform is clear: the sizeable power packs and the capacitors can be accommodated in a ship. This would allow the US Navy, at least, achieve its goal in providing precision fire support to ranges in excess of 500 km (Fair 2007).

So, let us take a look at the linear motor. The basic principles of a linear motor relies on the formation of the an electromagnetic force called the 'Lorentz force'. A basic schematic of a linear motor (that is applied to a railgun concept is shown in Fig. 3.4). Essentially, as a current flows through a wire, it generates an electromagnetic field (shown by the dashed circles in Fig. 3.4). Negatively charged electrons flow from the negative pole to the positive pole but the convention is commonly to represent current flowing from positive to negative and so I have done the same! The direction of the magnetic field around the wire is highlighted by the famous 'right-hand-rule' where you give a 'thumbs up' with the direction of your thumb showing the direction of the current flow and the way in which your fingers curl indicates the magnetic field flow direction (assuming positive to negative flow of current). The Lorentz force is applied to a charge particle moving through a magnetic field and therefore as the current flows through the projectile a force is applied as shown. The force direction is given by the 'other' right-hand-rule where your thumb, index finger and middle finger are orientated at right angles and provide directions according to the following:

- Your thumb indicates the direction of the current;

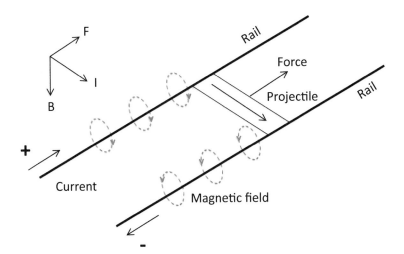

**Fig. 3.4** A schematic of a typical railgun arrangement; with directions shown for force (F), current flow in the projectile (I) and the magnetic field direction (B)

- Your index finger indicates the direction of the magnetic field, which in the example shown below is acting clockwise on the positive rail, or down;
- Your middle finger indicates to the direction of the resultant force.

There are two major drawbacks using railguns to get to very high velocities. The first being the power requirements. Enormous amounts of current are required to accelerate the projectile and usually this is achieved through banks of capacitors that allow for the delivery of current over short time scales. The other potential drawback is the fact that large temperatures are produced due to the high currents that are required. This was a major problem for some of the developments that occurred in the US in the 1990s and replacing the then-used copper rails every second or third firing does not lend itself to a practical system. The reason for this problem was due to small imperfection in the rails leading to localised extreme heating. This was overcome by injecting electroconductive liquid in the gaps between the armature and rails to provide a lubricant surface along which the armature can slide. This has been shown to be quite effective (Drobyshevski et al. 1999). Alternatively, high-conducting refractory materials would be required. That is, materials that have a high melting temperature such as tungsten or tantalum.

The projectile package has evolved into an armature (that is either C-shaped, or saddle-shaped), the design of which is optimised to minimise current density fluctuations (Satapathy et al. 2007), which in turn is attached to the penetrator payload (which would comprise of a dense penetrator, which itself would be saboted. There is nothing unconventional about using a sabot, as we have seen, and this is a common way of protecting the gun barrel and allowing the projectile to travel up the gun safely and securely. Both the sabot and the armature would be discarded after the projectile package leaves the gun barrel.

Another approach could be used is the coil gun. Here the projectile is electromagnetically propelled with successive energising of sequential electromagnets (see Fig. 3.5). This is the approach that has been touted for the development of the Space Gun where a gun would be built into the mountain to allow for a gradual acceleration

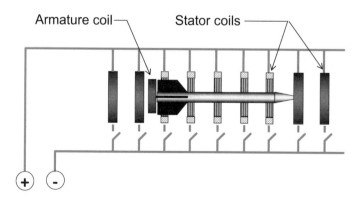

**Fig. 3.5** The coil gun

to high velocities. This was initially proposed in the 1937 fictional novel by Dr E F Northrop of Princeton University, written under a pseudonym (Pseudoman 1937). Of course, it is probably worth mentioning here the most famous space gun of all, that being the gun represented in the 1865 novel "From the Earth to the Moon" by Jules Verne where a bullet-shaped projectile is launched by 180,000 kg of gun cotton, from a cannon excavated into the ground.

The main problem with space launched projectiles is the immense aerodynamic resistance that a projectile would be subjected to and therefore a land-launched projectile would have to be supplemented with some type of rocket booster to carry the projectile into orbit. This is not to mention that firing humans in such a gun would be no fun and Verne's occupants would have certainly been squashed in the process, purely by the acceleration. You will recall that acceleration is the change of velocity with time and as the acceleration is increased forces are applied to the body called g-forces. High g-forces are quite destructive to our body tissue. Anything above 5-g, becomes uncomfortable. Trained, healthy and fit, pilots can maintain 10 g for a few seconds whilst wearing appropriate pressure suits and it has been known for humans to be decelerated at 46 g for a few seconds and survive.[7] However, even guns that achieve modest velocities will accelerate their projectiles to between 40,000 and 80,000 g—far in excess of survivable limits. So in summary, Verne's passengers would have been turned to 'mush' in the first several seconds of launch!

Nevertheless, it is worth noting here that Gerald Bull in the 1980s worked on the High-Altitude Research Programme and was able to get a projectile into 'almost' Low Earth Orbit. Note that during that time, satellites were about the size of a double decker bus. However, nowadays we find that satellites are relatively small packages and there has been extensive developments in what we call CubeSats. These are very small satellites indeed and often smaller than a typical artillery shell. So, it stands to reason that it is not beyond the realm of possibility that we could design a gun and a suitable propellant and a suitable rocket booster that we could attach to the satellite that could launch a CubeSat into outer space. It is also worth noting that these types of guns have been considered as being suitable launch methods from the Moon. That is, if mining the Moon becomes feasible. It is well known that there are numerous natural resources available on the Moon hitherto untouched. And so, when the technology allows, there would be an advantage of being able to mine the Moon and transfer those minerals back to Earth. As crazy as it may sound, these types of launch facilities could be used to fire back to Earth (from the Moon) large quantities of these precious minerals.

---

[7]This notable record was achieved by Capt John Stapp (1910–1999) who rode the Northrop "Gee Whiz" decelerator sled which was accelerated to up to 200 mph before a braking unit was applied to bring the sled to halt, thereby subjecting the passenger to up to 50 g in deceleration.

## 3.8 Electrothermal Guns

Additionally, a promising technology that could be integrated into existing gun designs relies on the integration of electrical energy with the energetic propellants. This provides the ability to modify the pressure time curve within the chamber. Electrothermal guns (or electro-thermal-chemical guns) have been the subject of intensive research efforts particularly in the US for a couple of decades. In their very simplest form, a high energy plasma is used in conjunction with a low molecular weight fluid to heat that fluid to a very high temperature such that the pressure within the chamber is dramatically increased. The resulting pressure drives the projectile forward. However, this would not give the velocity that would be required from a tank round. A more evolved approach is to use a plasma injected directly into the propellant to give a consistent ignition of the propellant granules.

A plasma is one of the four states of matter (after solid, liquid and gas). Essentially a plasma is a gas where some of the electrons have been stripped from the orbits of atoms. Therefore, positively charged ions swim in a sea of dissociated electrons. So, the gas is highly charged and can be conductive and highly influenced by magnetic fields. The addition of the plasma to the propellant leads to a reliable ignition of the energetic propellant and a more consistent burn. This leads to a number of benefits in terms of velocity, accuracy and consistency. With conventional propellants we find there are variabilities that arise in the pressure-time curve in the chamber due to a number of factors. This can be even as simple as the temperature of the ammunition before it is loaded into the chamber. It is well known that temperature plays a large role on muzzle velocity as any shooter would know. In fact, it is quite common that for someone who is conducting a large number of shots that the velocity of the round at the beginning of the day would be slower than the velocity of the round at the end of the day. The reason for this is that the temperature in the gun steel is conducting into the cartridge and thereby adding heat energy to the propellant. This increases the velocity. With electrothermal (ET) guns, we find that it provides a 'hot round velocity' across a broad range of temperatures (Dyvik et al. 2007). And, for tank guns being used in the Arctic, for example, this is particularly important. Furthermore, another advantage of ET technology is that it is it is claimed that accuracy is improved. This is probably due to a more consistent propellant ignition and burn. However, probably the most attractive advantage is the possibility of increasing the velocity of the projectile by virtue of tweaking the pressure time curve as seen in the chamber. This is achieved by the plasma, that is injected into the chamber, adding heat energy to the existing propellant and thereby providing for a longer time for the high pressure to be acting on the base of the projectile. Nevertheless, it is unlikely that we will see large increases in muzzle velocity that are entirely possible with railguns and coilguns.

## 3.9  Explosively Formed Projectiles

Another way of achieving high velocity in a projectile is by developing an explosively-formed projectile. These devices are able to propel projectiles at high velocity by virtue the fact that explosive energy is directed to the acceleration of a metallic dish that is inverted and compressed into a projectile. The explosive force that is required means that these types of weapons are not re-usable. In the strictest sense of the definition, these are not really guns per se but more like shaped charge devices. The fundamental difference to a shaped charge device is that instead of a conical liner being deformed into a jet, a relatively shallow dish is formed into a slug or projectile. The nature and size of the projectile can be optimised for a particular application by the use of different explosives, charge diameters, various 'wave-shaping' techniques, case and dish materials. The dish is often made of a relatively soft material to ensure that it deforms into an appropriate projectile-like shape. Relatively dense metals such as steel ($\rho = 7850$ kg/m$^3$), iron ($\rho = 7870$ kg/m$^3$), copper ($\rho = 8930$ kg/m$^3$), and, more recently, tantalum ($\rho = 16{,}690$ kg/m$^3$) are used to ensure effective penetrative performance, especially in the lower part of what is known as the 'hydrodynamic regime' (2–3 km/s). More on that later.

The discovery of the shaped charge effect is widely attributed to Franz Xaver von Badaar in 1792. Von Badaar (1765–1841) was a German Catholic philosopher, theologian, physician and mining engineer, to add to his talents! He advocated for the use of a hollow cavity at the front of an explosive composition (black powder) that would allow the explosive energy to be focussed. Black powder, however, does not detonate and so although the concept was correct, the true effect of an explosive charge was not realised until the works of Max von Foerster in 1883, who was Chief of the Nitrocellulose factory of Wollf & Co., in Walsrode, Germany at that time. Arguably, von Foerster was the true discoverer of the modern shaped-charge effect (Kennedy 1990).

The propulsive effect of explosive materials was rediscovered by Dr Charles Munroe (1849–1938) of the Naval Torpedo Station, Newport, Rhode Island in 1888. In one of Munroe's famous experiments, Munroe detonated blocks of explosive in contact with a steel plate. The explosive charge had the initials USN (United States Navy) inscribed on the charge which were then transferred to the steel plate by virtue of the detonation of the explosive charge. This has become known as explosive engraving (Walters 2008). Munroe, (whose name is often associated with shaped charge jets or 'Munroe jets') was the first to develop the lined cavity concept by defeating a large steel target with a tin can! According to Munroe, this was achieved by *'tying the sticks of dynamite around a tin can, the open mouth of the latter being placed downward'* (Munroe 1900)—see Fig. 3.6.

Munroe jets are fast moving super-plastic solid materials, now usually formed from copper conical liners in anti-tank guided missiles, torpedoes, cruise missiles and the like. However explosively-formed projectiles can trace their history back to 1936 when Professor R W Wood reported on the death of a young women inspecting her 'house furnace' to see if the fire was burning properly (Wood 1936). According

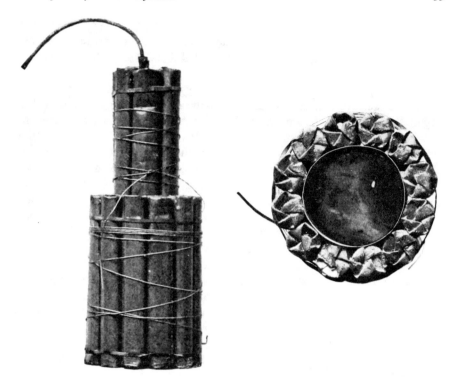

**Fig. 3.6** Munroe's hollow dynamite cartridge: elevation (left) and view from below (right); after (Munroe 1900)

to Wood's report, the young lady was "was struck by a small particle of metal which flew out of the fire and penetrated the breast bone, slitting a large artery and causing death in 2 or 3 min from internal haemorrhage." It turned out that the culprit was part of a detonator that had been delivered by mistake along with the coal for the furnace. At that time, the detonators that were used in coal mines for blasting were around 40 mm in length and 5 mm in diameter and formed from thin sheets of copper. The head of the detonator comprised of a concave component of copper. The mercury fulminate charge (usually detonated by an electrically heated wire) was set-off by the fire in the furnace resulting in the detonator's head being projected towards the young lady, who was opening the furnace door at the unfortunate time! So, infamously, she became the first victim of an explosively formed projectile. Wood proceeded to report all types of experiments that showed that the denotator heads were effectively turned inside out as they were projected at a high velocity (6000 ft/s, 1830 m/s) by the mercury fulminate composition—similar to our modern understanding of how modern explosively formed projectiles are produced.

It is entirely possible that Wood's paper could have prompted a Hungarian army officer, by the name of Misznay, to develop this observed phenomenon into a weapon during World War II. Thus, the Misznay-Schardin plate was invented. It is not clear how Professor Hubert Schardin contributed to the weapon design however it is thought that Schardin once visited Misznay in Hungry and on his return wrote a technical paper describing the Misznay-Schardin effect. And so the name stuck! (Kennedy 1990)

Unlike shaped-charge warheads, an explosively formed projectile is insensitive to stand-off distance (the distance between the warhead at initial slug formation and the target). Hence, they can be used in a wide variety of applications such as mines (for example the M70 Remote Anti-Armour Mine (RAAM) and the Yugoslav TMRP-6 anti-tank mine). It is used in top-attack sub-munitions such as those available in the M898 Sense and Destroy Armour (SADARM) projectile and guided weapons. They have been developed as top attack munitions simply because on armoured vehicles, that is where the armour has been the thinnest. They have also been extensively used in improvised explosive devices. These were particularly prevalent in Iraq and Afghanistan during the conflicts that occurred there from 2003 onwards. There are some slight nuances with the manufacture of explosively formed projectile munitions and these are generally around the manufacture of the liner and how it makes contact with the explosive. Generally speaking, the liner is made from a ductile material that is able to rapidly form into a slug. Therefore, attention is frequently given to the microstructure of the liner material to ensure uniform stretching. The other important factor is the quality of the explosive fill. Explosives can be quite unpredictable in their function when they contain holes or voids. These can lead to an inconsistent detonation. As the explosive is detonated, either through electrical ignition, or by some other mechanism, a fast moving detonation wave forms and makes contact with the liner, causing it to accelerate. In some ways one could argue that in this case we are using chemical energy, that is in the explosive, to accelerate a projectile. This is identical to a gun launched projectile where we use chemical energy, that is in the propellant, to accelerate a projectile. The notable difference here of course is that with an explosively formed munition there is no gun barrel.

## 3.10   Metal Storm

As I close this Chapter, I cannot ignore an innovative Australian invention that can be regarded as a 'fast gun'. Well, technically this invention is not a fast gun in the same way that we have been discussing fast guns. However, given that this weapon holds the World record for the fastest rate of projectile fire it is worth a mention here. Metal storm is a weapon system that is like no other. It is different to a conventional gun system and really breaks the mould of conventional gun design. So how does it work? Instead of a magazine that feeds the ammunition assembly individually into a chamber, Metal storm loads projectiles in series along a gun barrel. The main advantage here is that it is possible to remove the mechanical firing mechanisms for the gun that feed individual projectiles for firing. This does mean that an enormous rate of fire is quite possible by virtue of housing a 'line' of projectiles. This type of gun is called a superpose gun. On initiation, which occurs electronically from discrete initiation points along the length of the barrel, the projectile is accelerated towards the muzzle. Simultaneously, the expanding propellant gases pushes back on the projectiles seated behind the initiated projectile causing a push against a sealing band. Of course, it is worth noting that in 1792 a farmer by the name of Joseph Gaston Chambers presented a similar device to the US War Department. Chambers' invention was a flint-lock musket that had the capacity to fire a line of projectiles at 20 rounds-per-minute. This was a muzzle loaded weapon. The novelty in Chambers' design was that he used projectiles containing a small hollow cylinder that allowed for the burning of the black powder to ignite subsequent charges. Black powder burns relatively slowly, and by correctly positioning each projectile and charge mass in relation to one another a suitable volley of projectiles could be achieved. Unfortunately for Chambers, his early design was somewhat unreliable and therefore the US war Department was not willing to invest in development or wholesale purchase. However, by 1812, when the US declared war with Great Britain, Chambers had developed his invention. The musket had evolved into a seven-barrelled rifle. This volley-gun was able to fire 224 projectiles at a rate of 120 rounds-per-minute. Subsequently, the US Navy adopted the new gun and acquired fifty-three of the seven-barrelled swivels that would be mounted onto warships. Eventually the technology was abandoned due to widespread unreliability.

The main advantage of such a rapid-fire weapon is that it has the ability to pummel targets. That is, if you could create damage with your first projectile, and a fast-moving projectile follows through, then it is possible to achieve substantial levels of damage to even the most hardened targets. The other advantage with this type of technology is that it is well suited to be applied two aerial denial weapon systems. These are weapon systems that can defeat incoming ballistic missiles. However unfortunately even though Metal Storm was developed in the 1990s, commercialisation of the idea never really took off. In fact, the company Metal Storm Limited entered voluntary administration in 2012. Nevertheless, in late 2015 an Australian based defence research and development company called DefendTex took up the mantle and currently holds the IP with a view to develop the system. No weapons are currently deployed and in use, however.

# References

Barber JP (1972) The acceleration of macroparticles and a hypervelocity electromagnetic accelerator. Australian National University, Canberra

Charters AC (1995) The early years of aerodynamics ranges, light-gas guns, and high-velocity impact. Int J Impact Eng 17(1):151–182. https://doi.org/10.1016/0734-743X(95)99844-H

Chhabildas LC, Dunn JE, Reinhart WD, Miller JM (1993) An impact technique to accelerate flier plates to velocities over 12 km/s. Int J Impact Eng 14(1):121–132. https://doi.org/10.1016/0734-743X(93)90014-X

Chhabildas LC, Kmetyk LN, Reinhart WD, Hall CA (1995) Enhanced hypervelocity launcher—capabilities to 16 km/s. Int J Impact Eng 17(1):183–194. https://doi.org/10.1016/0734-743X(95)99845-I

Crozier W (1959) High velocity gun. US Patent 2,872,846

Drobyshevski EM, Kolesnikova EN, Yuferev VS (1999) Calculating the liquid film effect on solid armature rail-gun launching. IEEE Trans Magn 35(1):53–58. https://doi.org/10.1109/20.738375

Dyvik J, Herbig J, Appleton R, Reilly JO, Shin J (2007) Recent activities in electrothermal chemical launcher technologies at BAE systems. IEEE Trans Magn 43(1):303–307. https://doi.org/10.1109/TMAG.2006.887707

Fair HD (2007) Progress in electromagnetic launch science and technology. IEEE Trans Magn 43(1):93–98. https://doi.org/10.1109/TMAG.2006.887596

Glenn LA (1990) Design limitations on ultra-high velocity projectile launchers. Int J Impact Eng 10(1):185–196. https://doi.org/10.1016/0734-743X(90)90058-4

Hildebrand M, Huneault J, Loiseau J, Higgins AJ (2017) Down-bore two-laser heterodyne velocimetry of an implosion-driven hypervelocity launcher. In: AIP conference proceedings, vol 1793, no (1), p 160009. https://doi.org/10.1063/1.4971749

Kennedy DR (1990) The history of the shaped charge effect: the first 100 years. Los Alamos National Laboratory, Los Alamos, New Mexico

Munroe CE (1900) The Applications of Explosives II. Popular Sci Monthly 56(February):444–455

Ogorkiewicz RM (1991a) Technology of tanks I, vol 1. Jane's Information Group, Coulsdon, Surrey, United Kingdom

Ogorkiewicz RM (1991b) Technology of tanks II, vol 2. Jane's Information Group, Coulsdon, Surrey, United Kingdom

Pseudoman A (1937) Zero to eighty. Princeton University Press, Princeton, New Jersey

Satapathy S, Watt T, Persad C (2007) Effect of geometry change on armature behavior. IEEE Trans Magn 43(1):408–412. https://doi.org/10.1109/TMAG.2006.887683

Swift HF (2005) Light-gas gun technology: a historical perspective. In: Chhabildas LC, Davison L, Horie Y (eds) High-Pressure Shock Compression of Solids VIII: the science and technology of high-velocity impact. Springer Berlin Heidelberg, Berlin, Heidelberg, pp 1–35. https://doi.org/10.1007/3-540-27168-6_1

Walters W (2008) A brief history of shaped charges. In the proceedings of the 24th International Symposium on Ballistics, New Orleans, Louisiana, 22–26 Sept 2008

Watson JD (1970) High-Velocity explosively driven guns. Physics International Corp, San Leandro, CA, United States, Washington, United States

Whipple FL (1947) Meteorites and space travel. Astron J 52 (1161)

Wood RW (1936) Optical and physical effects of high explosives. Proc Roy Soc London A: Math Phys Eng Sci 157(891):249–261. https://doi.org/10.1098/rspa.1936.0191

# Chapter 4
# Big Guns

Big guns have their own unique challenges. They must contain the enormous pressures due to a large and heavy projectile being fired. They also present significant logistical burdens on the owners as heavy masses of steel need to be positioned and aimed with millimetre-accuracy. Some of these guns have come to be known as superguns and recent examples include Gerald Bull's Supergun that was almost built for Saddam Hussein, which some say led to Bull's death. However, before Bull, there were some very big guns and inventive minds. So, let us firstly take a look at these as we review the 'big guns'.

## 4.1 Medieval Siege Weapons

Many early bombards were enormous. The reason was simple: if you want to destroy large, heavy fortifications then you need to use large and heavy projectiles. One of the challenges of producing large guns was dealing with the huge stresses that arose during firing. Early bombards launched heavy projectiles—some as heavy as 300—400 lb, or 136—182 kg per shot. To keep breech pressures within safe limits the amount of black powder that was used was roughly one-eighth of the mass of the projectile for the largest examples (Davies et al. 2019). How the early gunners arrived at this ratio, no one knows for sure. Increasing the amount of propellant increased the risk of catastrophic failure, which would have almost certainly led to the deaths of the gunners. Small guns had been cast from bronze since the 1300s, but this was not without its challenges. Early castings had the potential to be porous, which would prove to be a weakness. Casting was a highly technical craft and expensive too. A much better way was to forge-weld iron, a technique well-known by medieval blacksmiths. Forge welding occurs when two metal parts are heated to a high temperature, so that the metal glowed, and mechanically forced together to create a join. It was effectively a mechanical welding process. Historically, this would have been

© Springer Nature Switzerland AG 2021
P. J. Hazell, *The Story of the Gun*, Springer Praxis Books,
https://doi.org/10.1007/978-3-030-73652-1_4

achieved by a blacksmith, some brute force and a hammer and was a common process in medieval times.

Even though medieval blacksmiths knew the methodology, the scientific reasoning for the join would have almost certainly alluded them. The brute force combined with the elevated temperature caused the atoms to literally 'drift' from one part to the next. This is a process that is referred to as 'diffusion' and it is way that is used to establish a resilient join. It is the same process that was developed to produce a Damascus sword which was made from multiple thin layers of iron. In this example, wrought-iron bars would have been hammered until thin, doubled-backed on themselves and then rehammered to produce a homogeneous (consistent) structure. The larger the number of times this process was repeated, the tougher the resulting sword blade.

The advantage of this technique was that the join was not only strong but tough. That is, it was resistant to cracking. However, the engineering required to carry out the process was quite demanding, even a reasonable challenge for today. For example, iron would need to be heated to temperatures of more than 1000 degrees centigrade. Further, these big, heavy structures, weighing in at many tonnes in some cases (see Table 4.1), were very difficult and risky to manoeuvre—especially when hot. That is why the manufacturing of these large weapons was very expensive.

For large iron siege weapons, the barrel was made by forming long iron staves which were positioned around a carefully crafted and ideally, perfectly round wooden log. These staves were forced into place around the log by heating up iron hoops (causing them to expand), slipping over the staves and allowing the hoops to cool thereby locking the iron staves in place. The wooden log was then burnt away. The purpose of the staves was to provide a smooth surface in which the projectile would travel. The staves themselves may have been subjected to the diffusion welding process however the join would not have been perfect. In fact, it is debateable as

| Dimensions | Mons Meg | Dulle Griet | Thanjavur |
|---|---|---|---|
| Origin | Belgium | Belgium | India |
| Year of manufacture | 1449 | 1430–1452 | 1600–1645 AD |
| Bore, chase (m) | 0.51 | 0.64 | 0.63 |
| Bore, chamber (m) | 0.22 | 0.26 | – |
| Length, chase (m) | 2.81 | 3.39 | 7.20 |
| Length, chamber (m) | 1.13 | 1.38 | – |
| Length, overall (m) | 4.06 | 5.03 | 7.51 |
| Final mass (kg) | 6040 | – | – |
| Current location | Edinburgh, UK | Ghent, Belgium | Thanjavur, India |

**Table 4.1** Examples of siege weapons made from iron (data from (Finlayson 1948; Balasubramaniam et al. 2004; Smith 1985)

to whether that would have even been necessary. To achieve a continuous cohesive joint would require a heated vacuum-press, which was well outside the technology of even the most talented medieval blacksmith!

One of the most famous of the bombards is Mons Meg—mainly as it is still on display at Edinburgh Castle. Mons Meg was classed as a heavy siege weapon that was built in 1449 with the purpose of destroying thick-walled structures. One of the more infamous stories about the bombard was that the first cannonball fired at the keep during the siege of Threave Castle in 1455 passed straight through the wall and severed the hand of Margaret Douglas as she was raising a glass of wine to her lips (Lewtas et al. 2016). Incredibly this bombard had a calibre of 510 mm and could fire a 490 mm diameter stone projectile to between 300–400 m/s (see Table 4.1 and Fig. 4.1). However, modern simulations have concluded that its cannonball would fail to break through 1.0 m of a castle wall, let alone the 3.0 m thick keep walls of Threave Castle (Lewtas et al. 2016). Perhaps poor Margaret Douglas was looking through the window of the keep just as the projectile crashed into a wall section nearby and was injured by a fragment of the cannonball. Perhaps it was simply a myth. Who knows? Margaret Douglas clearly survived the incident and died in March 1578. She was buried at Westminster Abbey and her tomb is there today—displaying both hands intact.

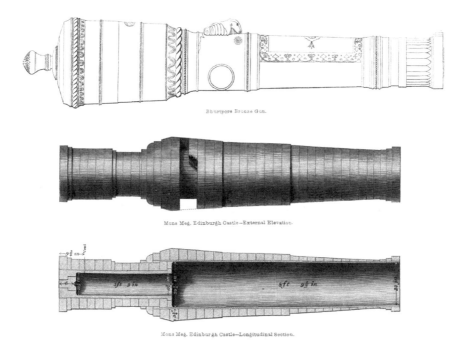

Bhurtpore Bronze Gun.

Mons Meg, Edinburgh Castle—External Elevation.

Mons Meg, Edinburgh Castle—Longitudinal Section.

**Fig. 4.1** Mons Meg (bottom) along with the Bhurtpore bronze gun after (Mallet 1856). The missing rings on Mons Meg (centre) was due to an incident in 1682 when the gun fired a salute. The iron rings were blown off (Johnson 1991)

In 1682 Mons Meg catastrophically failed during a ceremonial firing. The failure occurred at the weak point in the barrel construction where the wall thickness dropped markedly as the powder chamber transitioned to the barrel. The exact cause of the failure is unknown however it was clear that powder formulations were improving all the time during the life span of the gun. The increased pressure produced from a 'modern' formula may have contributed to Mons Meg's demise.

Bombards were a very fifteenth century invention. The Dardanelles gun was another such gun, which had a 25-inch calibre (635 mm) making it one of the largest working guns in history, in terms of calibre. Despite being built in 1464, it saw action as late as 1807 when the Turks used it against the British Fleet to hurl 630 mm stone balls at the tall ships. With an operational life of over 340 years, it may well hold the record for the longest-serving piece of moveable military equipment.

The Dardanelles Gun was made from cast bronze. This was presented to Queen Victoria in 1868. It was a one of a family of eighteen bronze guns in existence at that time. Even though bronze was more expensive than iron during medieval times, it had the advantage that it would show signs of strain before failure. Indeed, medieval wrought iron would show no evidence of strain before failing and therefore would give no warning if a gun was about to burst. However, bronze is heavy. It has a density of 8730 kg/m$^3$, about 11% higher than iron. Therefore, bronze guns tended to be heavy structures that were very difficult to move. Bronze is also softer than iron and so damage to the internal surfaces of the barrel would have been more evident. This would not have been a massive problem per se, however in time, wear to the internal surfaces of the barrel would have led to excessive leakage of the propelling gasses as they passed by the projectile in the barrel. Ultimately, this would diminish the velocity and therefore range. Projectile scores on the inside of the barrel of the Dardanelles Gun are evident today.

Probably the largest bombard to be manufactured in terms of calibre was the massive Tsar Cannon in Moscow with a bore diameter of 890 mm. It was cast in bronze by Andrei Chokhov in 1586 and weighed 40 tonnes. Alas, it was never fired in anger but serves as a nice "gate guard" at the Kremlin in Moscow along with some cannon balls, that were added in the 19th C. (Johnson 1991).

The pressures in these guns must have been enormous. The breech in the Dardanelles gun could hold up to 70 kg of black powder and huge pressures were generated as the propellant burnt and pushed against the very heavy stone projectile. It was largely understood at the time (quite correctly) that the wall thickness of the metal would provide suitable containment to the pressure generated by the black powder. Much of that understanding must have been learnt through trial and error. The theory for thick-walled pressure vessels (by which we understand how to contain pressures with a large thickness of metal) did not arrive until the work of Gabriel Lamé (1795–1870).

Lamé's work showed that the hoop stress (that is, the stress that the gun would 'see' around its circumference during firing) varied with the wall thickness of the barrel. Increasing the wall thickness resulted in a reduction of the hoop stress. His work also showed the stress acting longitudinally on the structure of the gun could be easily calculated. Until the 19th C. nobody had a good understanding of the strength of the structure required to stop a breech plug being accelerated toward the shooter during the gun firing. Therefore, it was no surprise that gun manufacturers gravitated towards muzzle loaders as they were a much safer bet!

## 4.2 Is Bigger Really Better?

Is bigger really better? This is a question that has plagued gun designers for over one and a half centuries. The question is: what is the advantage of increasing the bore size of a gun? There are two answers to that question. The most important reason on why you would want a large gun is simply down to the target effects that you want to achieve. However, there is also an often-forgotten physical principle at play and that is down to something called the ballistic coefficient. The ballistic coefficient (BC) of a projectile is defined as follows:

$$BC = \frac{m}{C_d A}$$

where $m$ is the mass of the projectile (in kg), $A$ is the cross-sectional area of the projectile (in $m^2$) and $C_d$ is something called the coefficient of drag—which is a measure of the drag forces acting on the projectile and is dimensionless. More on that later. The higher the ballistic coefficient of the projectile the easier it overcomes the air resistance during flight and, therefore, the longer its range. To have a high ballistic coefficient you need a projectile with a heavy mass, low cross-sectional area, and low drag coefficient. As gun scale increases, the projectile mass increases according to an approximate third-power law (due to an increase in volume as the volume is directly proportional to $d^3$, where d is the gun calibre). However, the drag on the projectile only increases by an approximate square-powered law (due to the increase in area). That is, as the diameter of the gun increases, the area of the projectile increases a greater amount due to the way we calculate area $= \pi d^2/4$. The area is directly proportional to the drag. That is exactly why a 9-inch pizza is a smidgeon over twice the size (in area) as a 6-inch pizza.[1] Therefore, the ballistic coefficient increases with gun calibre. There is also another advantage. Increasing the barrel diameter also reduces the peak accelerations experienced by the projectile. For 7-inch guns, peak accelerations are in the 40,000 g range whereas equivalent acceleration in the 16-inch guns are of the order of 20,000 g (Murphy and Bull 1966). So, yes, bigger is really better!

Range is important. However, there are many different factors at play when it comes to examining the range of a gun. These include the muzzle velocity of the projectile (which in turn is affected by the way the propellant burns), the size and shape of the projectile, the mass of the projectile, the drag on the projectile due to any facets that increase resistance, the elevation of the gun, and so on. Indeed, some

---

[1] And generally better value.

smaller guns can have higher ranges than some larger guns. For example, the French 380 mm/45 gun had a range of 42 km, longer than the Iowa's 16-inch /50 gun and similar to the Yamato's 460 mm/45 guns despite being smaller and of a similar era.

## 4.3  The Last of the BIG Muzzle-Loaders

A few hundred years after the reign of the bombards along came William Armstrong. Armstrong (1810–1900) was an English Engineer, started his career as a lawyer. This was not uncommon around this era—especially for the middle classes.[2] His natural flare and interest in engineering became apparent when, during his time as a solicitor, he developed a novel way to develop static electricity with his invention of the 'Armstrong Hydroelectric Machine'. Armstrong was tenacious and innovative and saw gaps in the armament market. He was knighted in Feb 1859 and by his death in 1900 the company that he founded, Armstrong Whitworth had become one of the largest armament manufacturers in the World; competing with Krupp of Germany, akin to Lockheed Martin, General Dynamics and Rheinmetall of today.

Around the time when Armstrong was developing new techniques to manufacture large guns, the newly formed Kingdom of Italy was developing its own Naval capability. It was recovering from a crushing defeat to the Austrian Navy. Ironically, the Austrians mainly won the battle by ramming their Italian opponents—not by gun fire. However, this did not stop Italy from pursuing new gun technologies. By 1866, Italy was forced to rebuild its fleet. This was for several reasons (see (Sullivan 1988), however probably the most important was due to souring Italo-French relations, due to Italy's refusal to support France in its war with Prussia. This resulted in Italy with two unfriendly northern neighbours, each possessing a superior fleet that could threaten the extensive Italian coastline. Furthermore, the Suez Canal was opened in 1869 and that promised to restore the strategic significance of the Mediterranean.

The building of the Italian Navy began with two iron-clad battleships: Caio Duilio and Enrico Dandolo (see Fig. 4.2). Italy had no capability to manufacture turreted gun systems—so the Italian Government turned to William Armstrong. They were fitted with the most powerful, and certainly the largest muzzle-loaded Naval guns ever used: Armstrong's 450 mm 100-tonne guns.[3]

However, one of the notable problems of muzzle-loaders in turreted systems was there was no clear method of checking that the round had fired before the new projectile was loaded. This was exactly the problem that the Royal Navy's HMS Thunderer (1872), launched March 12, faced. In 1879 a gun exploded killing eleven members of the crew and injuring thirty-five. The Thunderer was equipped with four

---

[2] A particular hero of mine, Bertram Hopkinson (1874–1918) started his career as a lawyer before turning his attention to engineering and eventually became Professor of Mechanisms and Applied Mechanics at Cambridge University at the age of 29.

[3] Despite the convention of naming guns by the mass of the projectile, Armstrong's 100 tonne gun broke that convention—the '100 tonnes' referred to the mass of the gun.

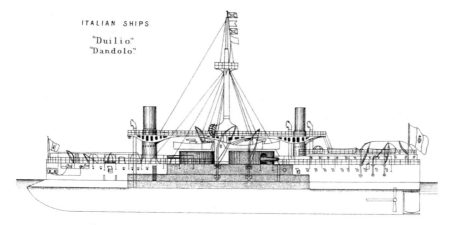

**Fig. 4.2** A line drawing of the Italian Iron-clads: Duilia and Dandolo showing the central location of the turrets housing Armstrong's powerful 100-tonne muzzle-loading cannons. Public domain

12-inch (305 mm) muzzle loaded guns and like many turreted ships with a muzzle-loaded gun, was loaded by depressing the elevation so that the gun was pointing at the deck. The charge and then projectile were subsequently rammed into place, from below deck in two operations (see Fig. 4.3). For one thing, it must have been a terrifying sight, seeing a massive 305-mm-calibre gun bearing down on you, as a loader, where you are sitting amongst kilograms of charge bags and explosive projectiles. Especially as this gun could send a projectile through 560 mm of iron combined with 150 mm of teak (King 1877).

It was clear that the gun explosion was caused due to the gun being double loaded. According to Admiral of the Fleet, E. H Seymour (1840–1929),

> Both turret guns were being fired simultaneously, and evidently one did not go off. It may seem hard to believe such a thing could happen and not be noticed, but from my own

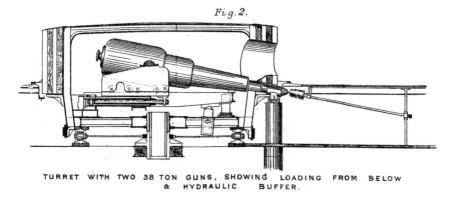

**Fig. 4.3** Thunderer's 38-tonne gun and its loading operation after (King 1877). Public domain

experience I understand it. The men in the turret often stopped their ears, and perhaps their eyes, at the moment of firing, and then instantly worked the run-in levers, and did not notice how much the guns had recoiled. This no doubt occurred. Both guns were at once reloaded, and the rammer's indicator, working by machinery, set fast and failed to show how far the new charge had gone. (Parkes 1990).

This forced the Royal Navy to have a rethink of its loading mechanism—moving to breach loading as opposed to a muzzle loading.

Despite muzzle loaders falling out of favour, the largest muzzle-loader ever built, previously installed on the Caio Duilio and Enrico Dandolo was installed at the Rinella battery on 12 January 1884. Armstrong's 100 tonne gun was the pinnacle of muzzle loaders—and there would not be a larger version in all of history. It could fire a 1 tonne shell over 8 miles and penetrate 15 inches (381 mm) of steel armour. To achieve this, it used a staggering amount of black powder: 200 kg. It also had one other advantage: it was rifled. This was a novelty for the time as most guns were smooth-bored despite the benefits of rifling being known for the previous two centuries (Blakely 1861). The rifling allowed for short projectiles to be spin-stabilised, the powerful gyroscopic forces acting on the projectile during flight kept the axis of the projectile true to the target. However, cutting grooves into a barrel to achieve this was thought to weaken the barrel and numerous attempts were carried out to achieve good rifling. This was overcome, in part, due to the increasing quality of steel and improved modes of gun manufacture, for which Armstrong was well versed.

The cost of firing the gun was enormous—roughly costing 100 lb sterling for every shot. In 1886 this was equivalent to the annual salary for over 6 men. Furthermore, it was estimated that due to the weight of the projectile and propellant charge that the barrel would only last 120 shots. This limited the peace-time firings to four per year.

On 5 May 1905 it was fired for the last time but was never fired in anger. A gun remains in place at Fort Rinella in Malta.

## 4.4  Big Bertha

At the turn of the twentieth Century the Germans developed a lead in big gun design because of the innovations of Alfred Krupp (1812–1887) and the developments of alloy steel in the nineteenth Century. The Krupp dynasty had a long history of armament manufacture. Alfred Krupp was given the nickname of "The Cannon King" and had, in 1851, presented a cast-steel cannon which became a sensation at London's Great Exhibition. Krupp's company also furnished the German military in both World Wars. However, it was in the first days of World War I that the German army revealed one of its newest secret weapons. This was a mobile 420 mm (16.5 inches) M-Gerät howitzer. This was a fort-smasher. Germans christened this gun "*Dicke Berta*" (big or fat Bertha) possibly in reference to Bertha von Krupp, owner of the firm that built the howitzers. German newspapers soon picked up the nickname that one can only assume would have been, well, uncomfortable and possibly offensive to

Mrs von Krupp. Eventually, "Big Bertha" was the name given to all German large-calibre artillery pieces until the K5 (E) railway gun was nicknamed "schlanke Bertha" (slender Bertha), a name most likely more appealing to the matriarch of the Krupp armament empire.[4] Basic data for the contenders are in Table 4.2 (Romanych and Rupp 2013).

Large mortars also became strategically important at that time. Mortars are short-range weapons, designed to lob large projectiles at high angles of elevation. Ideally, they are siege weapons and historically evolved from the times when a relatively small structure (the mortar) could be used to launch a heavy stone projectile to come crashing down on the enemy. Initially, large mortars were developed for installation along the German coastline. However, it was soon discovered that hitting a moving ship with mortar fire was near impossible. Therefore, attention was given to the development of mortars for attacking fortifications (Romanych and Rupp 2013). In 1893, the German Artillery Test Commission partnered with Krupp to develop the 30.5 cm or 12-inch mortar. Soon it was found that the German general staff embraced the idea of using heavy mortars to level land fortifications. The 30.5-cm design, with its high angle of elevation, could fire projectiles that would penetrate most, if not all permanent fortifications in Europe. However, being 30,000 kg, it was not particularly mobile and, in fact, had to be installed onto a plate, after the ground was levelled by hand and wooden foundation beams installed. This process was the predecessor to the modern base plate and bedding in procedure that the much smaller modern mortars employ. Roughly speaking it would take 12 h to get the mortar in place and prepare the mortar for action. Given that the range was limited (8.2 km) this was not ideal as it placed the mortar-line at risk of counter-attack once discovered.

This was a lesson that was not clearly learnt when the 42-cm Gamma-Gerät howitzer was introduced. This was the most powerful siege gun employed by the German army in World War I. It was also the least mobile. It had to be emplaced on a massive foundation made of steel and timber and set into a deep excavated pit that would have to be manually prepared. It would take 24 h to put in place under ideal conditions. There was, therefore, a need for a mobile howitzer. Since 1907 Krupp had been developing the concept of road-transportable weapons which ultimately led to the 42-cm M-Gerät howitzer. This was a mobile gun system. Its mass was less than one-third of the mass of the Gamma-Gerät howitzer and its range and payload was somewhat diminished. Nevertheless, the M-Gerät howitzer was regarded as the German army's ultimate siege weapon combining mobility and firepower into an effective fort killer. One shot from this Big Bertha hit a Belgium Brialmont fort and was able to penetrate an ammunition vault. The magazine exploded, demolishing the entire fort—killing 550 troops.

However, its maximum range was only 9.3 km. Further, World War I German siege guns had an unfortunate tell-tale signature. When the gun fired, a large plume of smoke topped by a ring was visible. This provided an opportunity for the allied

---

[4]The 28-cm Amiens gun, built in 1904 at the Krupp factory in Essen, Germany was given the nickname 'Little Bertha' or 'Baby Bertha'. The barrel and the roof structure now reside at the Australian War Memorial in Canberra.

**Table 4.2** German siege gun data from World War I, after (Romanych and Rupp 2013)

| | 28 cm L/12 i.R. & L/14 i.R. | 30.5 cm Beta-Gerät | 30.5 cm Beta-Gerät 09 | 30.5 cm Beta i.R. | 30.5 cm Beta-M-Gerät | 42 cm Gamma-Gerät | 42 cm M-Gerät |
|---|---|---|---|---|---|---|---|
| Mass (metric tonne) | 17.0 | 30.0 | 45.0 | 24.5 | 47.0 | 150.0 | 42.6 |
| Mass of heaviest projectile (kg) | 285 | 410 | 410 | 330 | 345 | 1160 | 800 |
| Rate of fire (rds/h) | 15 | 15 | 12 | 12 | 8 | 8 | 8 |
| Max. range (km) | 9.7 | 8.2 | 12.0 | 12.0 | 20.5 | 14.0 | 9.3 |
| Time to emplace (h) | 3–4 | 12 | 12 | 3–4 | 7–8 | 24 | 5–6 |

observers to spot the gun's location and attack. None of the German World War I siege artillery remain today.

Although the Big Bertha mortars were impressive, they were not the biggest in terms of calibre. During World War II, "Little David" was developed in the US which arguably was the largest mortar ever to be made with a calibre of 36 inches (914 mm). This was designed for test-firing aeriel bombs. Due to its massive size, the logistical impediment of getting it ready for firing and limited range (8.7 km) it was never used in anger as a weapon.

## 4.5   Long Range Artillery and the Paris Gun[5]

One of the more intriguing big-gun developments during World War I was that of the Paris Gun, managed by Professor Fritz Rausenberger (1868–1926), Chief of Artillery at Friedrich Krupp AG. These weapons were so-called as they were built for the sole purpose of bombarding Paris at a range of 120 km. During 1914, Rausenberger and his team had been investigating the possibility of long-range artillery. The German's artillery at that time had a maximum range of around 14 km and the German High Command had aspirations to bombard the English east coast port of Dover, a distance of 33 km from the French coast.[6] At that time, the guns available to Rausenberger were the Naval 30.5 cm L45[7] and a "new" 38 cm L45, known as the Langer Max. Also available was a Krupp experimental gun: a 35.5 cm L52.5 that had an anticipated range of 39 km. On 21 October 1914, Rausenberger began trials of a new low-drag projectile at Krupp's Meppen firing range using the experimental gun. After firing the first shot with a maximum propellant charge weight and a 43° elevation, Rausenberger's team waited anxiously for the location of the impact from the observers that were strategically located at the intended impact point. The report never came. Everything appeared to have been carried out normally and without incident and it was understood that the projectile had successfully left the gun barrel. So where had the shell landed? After several hours it was discovered that the shell had not impacted a site 39 km away (where the observers were located) but landed 49 km down-range from the battery. The shell weighed 535 kg, however it was inert. It landed in someone's garden which must have been a large shock to the property owner! Nobody was injured and there was relatively little damage. This unexpected but favourable result raised a fundamental question. How did the shell travel 49 km as opposed to the expected 39 km? After careful study it was deduced that the assumption of a constant air density along the trajectory of the shell's path was wrong. Air

---

[5]Much of this section of the book was gleaned from (Bull and Murphy 1988). This is a 'must-have' book for all students of 'big guns'.

[6]Although the separation of 33 km existed between Calais and Dover, it was anticipated that a gun with a range of 37 km was required to allow for variable weather conditions.

[7]The L45 denotes the length of the gun. In this case, the L45 gun was 45 calibres in length. The tradition of citing the length of the gun in terms of calibres carries on to this day.

density reduces with altitude. High-altitude trajectories meant that air resistance was largely reduced at the higher altitudes as the density reduced. When this was taken into account, the correct range was calculated. Remember, Rausenberger did not have the benefit of complex computational fluid dynamics calculations, high-speed computing or hypersonic wind tunnels. All of the calculations would have been done by hand. In some ways, it is amazing that they had the forethought to solve such problems. Nevertheless, this firing trial at Meppen made it clear that heavy, large calibre artillery projectiles with low-drag projectiles could achieve firing ranges well beyond those that were previously thought possible. This led Rausenberger's assistant Herr von Eberhard to speculate, using the newly understood atmospheric correction model that:

> Using the step-by-step atmospheric zone technique to give a correct approximation to varying air density, it was calculated that a 21 cm, low drag shell weighing 90 kg and launched with [a] velocity of 1500 m/s would have a maximum firing range of 100 km. (Bull and Murphy 1988).

This statement was the basis for the Paris Gun.

Dr von Eberhard suggested that to achieve a velocity of 1500 m/s with a 90 kg projectile, a saboted system could be employed—the first of its type. However, this was dismissed by Rausenberger in favour of using a very long barrel with a full-calibre projectile to achieve the same. Unfortunately, Krupp's rifling capability was limited to 18 m and the high velocity requirement coupled with a 90 kg projectile would result in high chamber pressures. Time was also short due to the pressures of war.

The solution was to repurpose some 35 cm L45 cannons that were being manu-factured by Krupp for the battle cruiser "Ersatz Freya". Rausenberger proposed to the Navy that he could expedite the long-range artillery development by using these barrels to insert a long 21 cm bore liner, that would be subsequently rifled. This would contain the pressure. To reach the desired velocity, smooth-bore barrel extensions were attached by means of a flange. After the long-range artillery project was over, Rausenberger argued that he could reconvert the 35 cm barrels to use on the battle cruisers. This never happened.

The 35 cm L45 gun 21 cm insert was made of three forged pieces and probably shrunk-fit in place. The chamber was bored to accept the Navy's 28 cm L45 cartridge. To achieve the desired velocity, a 6 m smooth bore extension was added taking the total barrel length to 28 m. However, this resulted in a predicted droop of 90 mm. Droop is proportional to the cube of the length and therefore adding relatively small sections onto the end of an existing system can lead to large increases in droop. To correct the droop a gantry was erected to support the barrel along its length. For an increased range, the length of the gun barrel is important as long as there are modifications to the propellant system. For most gun systems there is generally a positive pressure at the muzzle and that can be made use of by extending the length of the barrel. However, by changing the way in which the propellant burns and adjusting

| **Table 4.3** Barrel-chamber parameters of the Paris Gun, after (Bull and Murphy 1988) | | 35 cm L45 | 38 cm L45 |
|---|---|---|---|
| | Length of original barrel (m) | 15.8 | 17.1 |
| | Tube insert length (m) | 21.0 | 21.0 |
| | Chamber length (m) | 3.0 | 3.0 |
| | Chamber breech diameter (mm) | 357 | 357 |
| | Estimated chamber volume (cm$^3$) | 260,000 | 260,000 |
| | Smooth bore attachment lengths (m) | 6.0, 9.0, 12.0 | 6.0, 9.0, 12.0 |
| | Rifling twist (right-hand) | 1 in 35 | 1 in 35 |
| | Width of lands and grooves (mm) | 4 | 4 |
| | Depth of grooves (mm) | 3 | 3 |

what we call the 'all-burnt position,[8] it is possible to extract more internal energy out of the propellant that is transferred to kinetic energy of the projectile. This also has the advantage of reducing the acceleration of the projectile.

Between 2 and 3 Naval 38 cm L45 guns were also repurposed for the Paris Gun project. The two basic configurations of the Paris Gun design are listed below in Table 4.3. Note that the 21 m tube insert protruded from the end of the L 45 gun 'casings'. For maximum range (120–127 km), the 12 m smooth bore attachment was required.

The maximum length of the Paris Gun was 34 m. This comprised of the 21 m tube, a 12 m smooth bore extension and 1 m of breech ring. The huge length of the gun barrel coupled with a large mass of propellant (250 kg) meant that it could lob a shell to an altitude of 39 km before it came whistling down towards its target.

After all this effort the Paris Gun was largely seen as a failure. A total of 351 rounds inflicted an estimated casualty total of 256 dead and 620 wounded. The psychological effects of the weapon were largely limited too. Nevertheless, it was later to provide the inspiration for a space gun, with project HARP.

## 4.6  Railway Guns

The ability to move a gun into place is essential for artillery. In fact, most modern armies today rely on and maintain a towed or moveable artillery capability. However, horse-drawn weapons in the 1800s limited the mass of the gun that could be towed and therefore in the 1860s during the American Civil War new moveable railway guns were developed. The largest moveable weapon at that time was a mortar called

---

[8]That is, the position along the gun barrel at which the propellant has been completely consumed during deflagration.

**Fig. 4.4** The Dictator—one of the first 'railway guns' (public domain)

the 'Dictator'. All in, the Dictator weighed around 10 tonnes with a calibre of 330 mm (13 inch) a was simply used to 'lob' munitions into fortifications (see Fig. 4.4).

Railway guns became famous in World War I especially before the war became entrenched in a static combat. Most of the early railway guns were converted from naval guns and were large heavy structures. The barrels were, quite frankly, enormous. For example, the World War I era 28-cm Amiens gun barrel alone had a mass of 46 tonnes. The thickness of the steel was large so as to contain the high pressures during firing (remember Lame?). Further, the projectiles that were being launched were huge. The main advantage on the Western Front was that extensive use of the French rail network could be made without getting bogged down in the muddy fields. Further, lethal payloads could be quickly delivered before retreating to relative safety, well behind the front line.

Before the start of the First World War, British forces mounted 4.7-inch. guns on railway cars. This was done during the Second Boer War 1899–1902 with great success. During the First World War, the British deployed 9.2-inch, 12-inch and 14-inch naval guns on specially designed railway truck mountings. It was felt that heavier firepower was required to defeat the Hindenburg Line and so the Ordnance Committee invited the Elswick Ordnance Company and Vickers Ltd to submit designs for a railway gun based on designs of a howitzer barrel of 18-inches calibre and length of 35 calibres. However despite six being manufactured, none would contribute to the war effort in time (Magrath 2015). In 1940 one of the 18-inch Howitzers, was stationed at Bishopsbourne, Kent on the Elham to Canterbury line as a defence against possible German invasion, which of course never came.

As seen with the Paris gun, one of the challenges of long gun barrels is that they will tend to droop under their own weight. And increasing the length of the barrel even by a small amount can have a detrimental effect in that regard. Put it this way, increasing the length of a barrel by one quarter can lead to a doubling in the amount by which the barrel will droop at the muzzle end. Modern guns can accommodate the droop by using what is called a 'muzzle reference system'. This is a device that liaises with the fire control system as to the exact location of the end of the barrel (usually using some form of laser positioning system). However, 20th C. railway guns did not have that luxury. On the one hand, one could argue that this was largely irrelevant as these weapons were used to pound cities. Being off target be a few tens-of-metres was not a drawback.

Railway guns were also used in World War II. Arguably one of the most significant of the railway guns of this era was the 28 cm K5 (E),[9] which had an intended range of 50 km. The design was jointly carried out by both Krupp, learning from extensive experience with long-range artillery, and Hanomag of Hannover. This work started in 1934. Initially the gun was prone to barrel cracks which was due the deep rifling grooves (10 mm) that were employed. Why Krupp adopted 10-mm grooves is a mystery, given the Paris gun employed only 3 mm grooves (see Table 4.3). It was presumably so that the projectile was guaranteed to spin correctly. The standard shell was fluted with angled splines (see Fig. 4.5) that were fed into the deep grooves and it was necessary to index the projectile with the grooves in the barrel (Anon 1988). Later, 7 mm deep grooves were adopted which appeared to alleviate the barrel cracking problem. In total, 22 weapons of this type were made (Zaloga and Dennis 2016).

However, probably the most impressive gun at that time as the enormous 80-cm calibre K(E) Schwerer Gustav and its sister gun, Dora, so named after the wife of Erich Müller, the head of the Krupp design bureau. In the Spring of 1942, the Gustav became the largest gun to ever have been used in combat. These guns weighed 1350 tonnes, were four stories tall and their barrels were 30 m long. The guns were also expensive! Each gun was equivalent in cost to 25 Tiger tanks. Hitler was obsessed by large megalomaniac projects such as this but had not learned the lessons from the Paris gun of World War I. Even though it was a fearsome gun, its potency was questionable. Dora was first used in combat on 5 June 1942. During this time, the gun fired 15 rounds against three Soviet targets with little-to-no effect. There seem to be large explosions and dust kicked up from the impacts however only one round was suspected of hitting anything of note. The average distance that the round would miss the target was about 300 m (Zaloga and Dennis 2016). The Dora was equipped to fire two different types of projectiles: one was a 4.8 metric tonne high-explosive projectile, and the second round was a 7-metric tonne concrete penetrator for bunker-busting. Accuracy problems plagued the Dora and certainly for something like a concrete penetrating projectile this was a significant impediment. All in all on the Russian front, Dora fired 48 rounds totalling 360 metric tonnes in less than two weeks without disabling any of its primary targets (Zaloga and Dennis 2016).

---

[9]K(E) = 'Kanone Eisenbahn', which literally means 'cannon railway'.

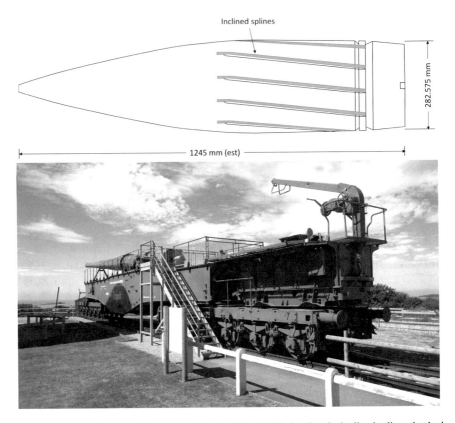

**Fig. 4.5**  Top: A schematic of the 'standard' shell of the K5 (E) showing the inclined splines that had to be indexed with the rifling. Adapted from (Anon 1988)). Bottom: the K5(E) 'Leopold'. *Source* Christian Randall

The Dora and the Gustav soon fell out of favour with the Army. Firstly, due to the huge size and weight, these guns had to rely on specially constructed railway tracks which were incredibly labour-intensive to lay down—requiring a crew of over a thousand men or conscript prisoners just to bring the gun towards its target. Secondly, the constructed railway tracks made an easy-to-follow path for Allied bomber crews making the guns easy prey. This was mitigated against by the clever use of phantom guns carefully deployed to distract the Allied bombers, however this again simply exhausted further manpower resources. Thirdly, it would take up to three and a half days to setup the gun and it could only fire fourteen rounds per day. Even with the large shells, the mass delivered to the target could not replicate a battery of smaller agile artillery pieces with a much higher rate of fire—such as the Czech Heavy howitzer model 25 that could fire a 42 kg shell at a rate of two-per-minute. Fourthly, and the final nail in the coffin, the A-4 ballistic missile, better known by its propaganda name the "V-2" was introduced which could deliver a 1-tonne explosive payload to 270 km, five-times the range of the 80 cm guns with less exposure and

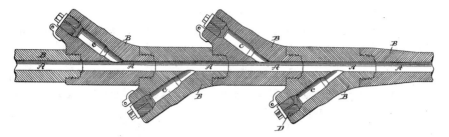

**Fig. 4.6** Haskell's multicharged gun from (Haskell 1892). Public domain

vulnerability to resources on the ground. Thus, eventually, Hitler's massive railway gun project was abandoned.

Other nations also toyed with railway guns. The Italian Defence Force played with a 120-millimetre 45 calibre gun system as well as a 152 millimetre 40 calibre weapon. The Soviet forces also developed some impressive railway weaponry including the mighty TM-1-14 which had a calibre of 356 millimetres and a gun tube length of 52 calibres. However, none of these matched the awesome Gustav or the Dora. Most were reengineered and redeployed Naval guns. Finland Japan Britain and the US all experimented with railway guns of different sizes but with limited success.

## 4.7 The German V3

A much more suitable solution for the German's appeared to lie in a concealable stationary massive gun with a single target in mind. This was the purpose of the V-3.[10] Engineering a gun to achieve a suitable muzzle velocity to realise an extended range was no easy task with conventional gun approaches. As previously noted, to achieve a very high velocity it is important to keep a sustained push behind the projectile. This is impossible to do with a conventional gun design as the pressure behind the projectile diminishes as the volume behind the projectile expands. However, if it were possible to have multiple stages, where propellant was ignited at various positions along the length of the gun barrel, then a sustained push was more probable.

In 1945, in a bombed-out munitions factory, the British discovered the secret plans for a Nazi terror weapon, the V-3. Made from sections, the V-3 was 160 m long— like Bull's proposed Supergun (see later). Parts of a prototype were also found by members of MI10, a British Army intelligence unit tasked to assess captured German war materiel. This gun comprised of multiple ignition stages for the propellant along the length of the gun. It almost certainly inspired by previous versions of a multi-stage gun that was developed towards the end of the 19th C by J R Haskell (Haskell 1881, 1892)—see Fig. 4.6.

---

[10]The 'V' stood for vengeance.

One of the major problems with this design was that hot gases from the burning propellant would accelerate past the projectile whilst in the gun barrel. These hot gases would cause premature ignition of the subsequent stages. This problem was solved by placing a sealing piston between the projectile and the initial propellant charge. This was successful in preventing the flash from the burning propellant from getting ahead of the projectile. Much of the work was carried out during the middle of 1944 when the Germans were on the back foot.

Although the main concept was technically sound, in practice the weapon system had two major downfalls. The first was the rate of fire. Literally tens of breeches had to be loaded with their own individual propellant charge for each successive firing. This would have been time consuming and problematic without an army of artilleryman to support single firing. The second problem was the fact that as it was a large structure, it was immovable and therefore susceptible to attack by the Allied forces, despite concealment. Existence of the weapon was highly classified as any indication of its location would ultimately have led to its doom. And that is exactly what happened.

## 4.8 Atomic Annie—280 mm Calibre

After World War II, it became apparent that there was a growing requirement for a gun that could fire an atomic shell. The challenge was ensuring high reliability of firing, guaranteeing excellent range as well as the built-in safety expected from the ammunition design. The end result was the 280 mm nuclear artillery weapon. The artillery system was dubbed the M65 atomic cannon or Atomic Annie. On 25 May 1953 at 8:30 am the atomic cannon fired its first and only nuclear shell at the Nevada test site. It was fired to a range of 11 km and within 8 s of firing, the nuclear payload detonated at a low altitude. The low altitude detonation was required to maximise the pressure of the shock due to constructive interference as the shock wave rebounded from the ground. Although the results were impressive, this was not the only nuclear shell that was fired from a gun. In fact, later in 1957 a nuclear AW 54 warhead was used with the Davy Crockett weapon system. The M-28 or M-29 Davy Crockett Weapon System was the tactical nuclear gun for firing the M-388 nuclear projectile. This was a recoilless smoothbore gun.

Atomic Annie was eventually deployed in Europe and in South Korea however the system had several drawbacks. The first was the concern for those who were firing the weapon and their proximity to the nuclear blast. This second concern was the way that it was transported and the logistic burden that it placed on defence forces. It was transported using two independent tractors and the drivers found it very difficult to negotiate around the paved roads of Europe; occasionally it was susceptible to tipping over. This gave it one more nickname and that was the 'Widow Maker'. Eventually around twenty of these were produced and today around seven are on display. The scientific challenges of such a weapon system are enormous as the shell itself would weigh 360 kg. And for such a heavy shell, enormous pressures would have been realised in the chamber. By all accounts, due to the high pressures and temperatures, the breech was at risk of welding shut during the firing.

## 4.9 The Space Gun

When J F Kennedy lay down the mantle on 12 September 1962 that the USA would put a man on the Moon before the decade was out it was an ambitious aspiration. This was particularly true as it was to be done with "new metal alloys, some of which have not yet been invented". Nevertheless, as we all know it was successfully achieved on 20 July 1969. Now of course, much thought had already been undertaken for the Moon landing before Kennedy's great "Moon speech" at Rice University but it set up the 1960s to be one of the most exciting decades for science and engineering.

Technically, the inventor of the space gun was Isaac Newton. In his book: *A Treatise of the System of the World*, Newton discussed a thought experiment where a cannon was placed on a very high mountain and shot a cannonball at a horizontal elevation (Newton 1731). If the velocity was low, then the projectile would simply fall back to Earth due to gravity pulling it down. However, if the velocity was sufficiently high then it would be possible to overcome the pull of Earth's gravity and escape. Of course, there are huge technological impediments that needed to be addressed, however some believed that these could be overcome.

Around the time of Kennedy's speech there was considerable interest in finding an accessible path to launch satellites into space and a gun seemed the obvious solution. The main challenges would be the high gun shocks and the acceleration that would be required to get a payload into orbit. Not an easy job! Of course, this was not a new idea as Newton had already raised the idea with his thought experiment. Jules Verne had also thought about it almost exactly a hundred years earlier in his 1865 novel "From the Earth to the Moon". In Verne's novel, a columbiad[11] gun is used to fire an aluminium projectile to the moon with the intent of delivering three passengers, who were comfortably sitting in the projectile, to the surface of the moon. Aluminium was the new "wonder metal" at the time that Verne wrote his classic novel, and it held great promise as a strong, lightweight material. Alas, although a tantalizing science-fiction possibility of getting people into space it is not possible simply because of the limited acceleration that the human body can withstand. Speed itself is not the issue and so if we can reach the velocity required to escape the earth's pull (11.2 km/s) slowly we can happily travel at that velocity without too much concern as along as we are protected from the elements. In fact, as you read this book you are currently whizzing around the sun at a staggering 30 km/s or 67,000 miles per hour.

Acceleration is measured in terms of a g force where $+1$ g is the due to the pull of the earth on our bodies. In short, it is the 1 g, that stops us floating off into outer-space. It is a force that we all are familiar with when the passenger airline takes of and pins

---

[11]"Columbiad" was the name given in the US to large-calibre, cast-iron, smoothbore, muzzle-loading cannons that were able to fire heavy projectiles.

us to our seats. Military pilots have to sustain the highest acceleration that can reach as high as 25 g during ejection from a fast-moving plane. However, this is for a very short duration and most people will lose consciousness with a sustained acceleration of 12 g. Now let us consider the accelerations in a gun barrel. In a 'conventional' Vernesque columbiad gun one could expect the accelerations to be many thousands of g (possibly as high as 100,000 g) and therefore all bets are off as far as human survivability are concerned (as we have discussed before). That is unless we extend the length of the gun and change the way in which we launch.

The other major obstacle for a space gun, is of course, the atmosphere. The atmosphere places a huge 'drag' force on the projectile as it passes through. And, things get worse as you increase the velocity of that projectile. In the transonic region, where the velocity of the projectile transitions from subsonic to supersonic, the drag forces increase substantially. Arguably the only way that this could be overcome with a gun-fired projectile is to give the projectile an extra push with an in-built rocket just as it reaches the maximum altitude.

### 4.9.1   Projects HARP and SHARP

A space gun was exactly the purpose of project HARP or the High-Altitude Research Program! By the 1960s two World Wars had left behind a legacy of large calibre, powerful guns that had the potential to fire projectiles huge distances and indeed into low Earth orbit. With some inspiration from the Paris Gun project of 1914–1918, project HARP began.

HARP used two 16-inch guns joined together to form one long barrel that was a staggering 85-calibres in length. This meant that the barrel was approximately 34.5 m long. The barrels were smooth-bored barrels and therefore conventional 16-inch Naval guns had their rifling removed so that they each had an internal diameter of 16.4 inch (Murphy and Bull 1966). New propellant designs were required to make uses of the extended barrel length. However, the main novelty of the system was using a specific type of projectile which was branded 'Martlet'. There were several evolutions of the Martlet projectiles culminating in a Martlet 4 that comprised of three stages of rocket motors that could carry a payload. The Martlet 4 had a launch mass of just over 900 kg and could theoretically place a 22 kg payload into an orbital trajectory. The first stage of the rocket would ignite at approximately 45 km with the second stage igniting at approximately 67 km and the final boost occurring at 325 km (Murphy and Bull 1966). Well and truly into orbit!

Of course, like all federally-funded projects, project HARP had to find broader appeal. One concept that was proposed was to design a projectile to obtain meteorological and geophysical data from gun launched experiments (Murphy et al. 1969, 1972). This work was carried out in the late 1960s however, it did not save HARP from the inevitable budget cuts. The rusted remains of the project are still on the island of Barbados.

One has to say that project HARP was far from being a complete failure. The harsh reality was that the US had committed to invest in rocketry and therefore gun launched projectiles fell out of favour. However, it is entirely possible that these systems could find a re-emergence. The reason for that is that in the 1960s satellites were likely to grow in size and eventually in the 1980s became quite sizeable. Nowadays the trend is to reduce the size of satellites to small payloads and CubeSats. Who knows? There may still be some mileage in the space gun.

There was one final 'push' to solve this problem in the 1980s. And that was the Super High-Altitude Research Project or SHARP. This project abandoned the use of propellant guns and used the well-established technology of two-stage light-gas guns, that were discussed in Chap. 3. This project was headed by Dr John Hunter and achieved velocities of around 3 km/s with 5 kg projectiles. However, the project was cancelled before the team at Lawrence Livermore National Laboratories was able to attempt a shot into space.

So, in summary: Almost 300 years after Newton proposed the thought experiment involving a space gun, we have no space gun.

## 4.10 Project Babylon (Saddam Hussein's Supergun)

And so we arrive at the Supergun. Put simply, this is the gun that got Gerald Bull (1928–1990) into a lot of trouble! You see Saddam Hussein wanted to build a 'Super Gun' that could target Israel from Iraq. Bull started working for the Iraqi leader during the Iran-Iraq war and the intention was to develop a 1000 mm calibre weapon that would have a barrel length of 156 m and buried into a secret location in a mountain north of Baghdad. This would have the capability of firing over 1000 kg projectiles great distances that would have potentially upset the power balance in the Middle East. Capable of being able to launch conventional as well as nuclear shells, the weapon would have given Hussein a distinct advantage over his neighbours.

Initially a 350 mm, test weapon was to be constructed with barrel pieces manufactured at different suppliers around the world and shipped to Iraq. That way, by using different suppliers, the thick-walled tubes could have easily passed off as tubes to make pressure vessels for the petrochemical industry. However, in April 1990 the UK's HM customs uncovered the plot and impounded eight crates containing some of the full-sized barrel sections—effectively stopping the project.

Sectioned barrels are quite unusual and almost never really used due to the high stresses in on the steel. However, Bull had some precedent with Project HARP and he was well versed with the 1918 Paris Gun that attacked the French capital from 120 km away.

Bull was assassinated on the 6th floor of his apartment building by five shots fired from a silenced automatic weapon. It was largely believed to be a state sponsored assassination, although the perpetrators have never been caught. He is buried in Montreal, Canada.

## 4.11  Howitzers

Modern large guns tend to be branded howitzers. The English word 'howitzer' originates from the Czech word *houfnice* which in turn is derived, from the word houf, "crowd". This suggests the original intention of the weapon was to fire against groups of enemies. In the Hussite Wars of the 1420s and 1430s, the Czechs used short barrelled houfnice cannons to fire at short distances into crowds of infantry. The word was rendered into German as 'aufeniz' in the earliest attested use in a document dating from 1440; later German renderings include 'haussnitz' and, eventually 'haubitze', from which derive the Scandinavian 'haubits', French 'obusier' and the Dutch word 'houwitser', which led to the English word 'howitzer' (Fisher 2020).

Up until around the end of the 17th C. artillery had been designed to be serve in one of two functions. Either there was an intention of high-angle fire (mortar) or horizontal fire (cannon). In 1693 a new piece of weaponry appeared in Holland that combined both attributes—the howitzer. Subsequently three basic types of artillery weapons existed, principally defined by barrel length which in turn was described in terms of a multiple of the calibre. So, a 39-calibre 105 mm gun would have a barrel length of 39 × 105 mm = 4.095 mm (or 4.095 m). The three types of weaponry with their respective lengths were as follows:

- Mortar—up to 12 calibres;
- Howitzer—12-30 calibres;
- Gun—over 30 calibres (Cleator 1967).

Such taxonomy has since been somewhat lost. Nowadays we define a howitzer as a weapon system that is designed to lob projectiles to large distances. Many howitzers have barrel lengths that are larger than 30 calibres in length with the most common barrel length bring 39 calibres in length for a 155-mm and there is work under way to extend that length further. Modern systems are designed for long range attack however in keeping with the original definition of a howitzer from the seventeenth century they can often be used in a direct-fire mode.

Most modern howitzers are in the 152-mm/155-mm calibre range. Typical examples include the FH70, the M198, and more recently the M 777. The M 777 is a relatively new version of the 155 mm howitzer. And it can deliver a 40-kg shell to the target at a range of between 30 and 35 km, when a rocket assisted projectile is used and around 25 km when the shell is unassisted by propulsion. What is unique about this weapon system is that unlike its previous counterpart the M 198, its mass is relatively low. It also has shorter trails. Modern warfare has become heavily dictated by logistics. This means with something like a 155 mm howitzer it is important to be able to transport it as far as possible whilst minimising the number of people required to achieve that. Reducing weight also has the benefit that the weapon can be lowered into position by helicopters and easier to transport by aeroplane. This is achieved as well by the fact that it uses titanium alloy in a large part of the structure. Titanium is a relatively lightweight material has a low density of 4450 kg per cubic metre, but of course, reasonably costly! However, it is the strength-to-weight ratio that makes titanium alloys attractive. Materials that have high strength-to-weight ratios are generally attractive for a range of engineering scenarios where extreme loads are expected. Furthermore, we know that titanium has very good corrosion resistant properties and so has not only been adopted by the aerospace industries but also the medical community too. In fact, it is not uncommon to find to titanium alloys in people who have shattered bones.

Howitzers are common-place today and are used in both the towed variety as well as the self-propelled version. The self-propelled versions give the advantage of providing protection under fire and being an independent mover. Most Armies prefer having both capabilities at their disposal.

## 4.12  Coriolis

One of the notable features of big guns is that generally they are designed to fire projectiles to enormous distances. I remembered being 'wowed' at the prospect of a firepower demonstration on Salisbury plain in England that a 155 mm howitzer to could fire a projectile and hit my office located around 40 km away at the UK Defence Academy (then called the Royal Military College of Science).

Large distances pose some problems in terms of accuracy and an impact to within a metre or two frequently does not matter for most artillery units. Missing your target by several hundred metres clearly would, however. Deviation during long range firings occurs due to a phenomenon called drift. An artillery shell fired over a distance of 20 km could have a drift imparted to it as much as 100 m depending on the gun's geographical position, direction of firing and time of flight. For ranges of less than 5 km the drift is generally less than the shot-to-shot variations at the point of impact whereas for small arms, the drift is deemed negligible and hardly measurable (Farrar and Leeming 1982).

One of the causes of drift in artillery shells is due to the Coriolis effect. It was not until 1835 that a full mathematical description of the Coriolis force was documented by the mathematician who gave this force its name, Gaspard-Gustave de Coriolis (1792–1843). Deflection of an object due to the Coriolis force is called the Coriolis effect. The Coriolis force occurs due to the spin of the Earth and more specifically the fact that points on the Earth's surface spin more quickly at the equator than they do at the poles, and this is due to a simple fact that the Earth is (very nearly) spherical. The Coriolis force also explains why cyclones are counterclockwise-rotating storms in the northern hemisphere, but rotate clockwise in the southern hemisphere.

In simple terms, at the equator, the surface of the Earth has a higher linear velocity than anywhere else on the Earth's surface. A point on the equator will have a linear velocity of 450 m/s that is moving from west to east and this velocity is less as we move towards either of the poles. So, if we had a gun that was located on the equator and we fired a projectile northward intending to hit a target due north then we would miss the target. Instead, the projectile would 'drift' toward the east. The reason for this is because the gun is located at a point on the Earth that is spinning

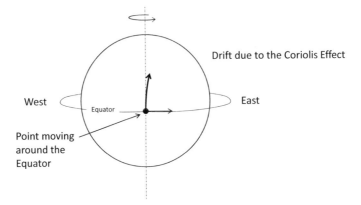

West                                                                    East

**Fig. 4.7** The spin of the Earth dictates the point of impact as a projectile is fired northward

faster than where our supposed target is. As the projectile is fired from the gun, the lateral momentum imparted by the Earth's spin at that point must be maintained but as the target is moving slower than the gun (from west to east) the projectile has a tendency to drift eastward (see Fig. 4.7). If we fired toward a target in the southern hemisphere, the projectile would also drift eastward (to the left as opposed to the right in the northern hemisphere). So, an identical effect is observed.

A similar effect is seen when firing from west to east and east to west at the same points of latitude. This due to what we call the Eötvös Effect and is named after a Hungarian nobleman and physicist Baron Roland von Eötvös (1848–1919). With this effect, it is observed that when we fire a projectile eastward, that is in the same direction as the rotation of the Earth, the Earth effectively rotates down and away from the trajectory so the projectile passes over the target (or at a higher point) and so would strike the Earth at some point beyond the target.

So, all in big guns have their challenges but they also have their advantages. Generally speaking, practicalities govern the size of a gun. That is, how it is going to be deployed, how easy it is to move, how resilient it is under firing, how much it costs through its whole life cycle, the number of people required to operate it and so on. In my view, unless there is sufficient ambition to bring back the space gun, it is unlikely that we will see gun systems growing in size as they did on the 19th C. Indeed, in the military space, as we have seen with the M 777, there is now a push to reduce the mass of these weapons systems. It remains to be seen as to whether or not there is enough ambition to bring back Gerald Bull's space gun. I for one, certainly hope so.

# References

Anon (1988) German explosive ordnance, artillery ammunition, rockets, launchers and missiles: World War Two. Weapons and Warfare Press, Bennington, USA

Balasubramaniam R, Saxena A, Anantharaman TR, Reguer S, Dillmann P (2004) A marvel of medieval Indian metallurgy: Thanjavur's forge-welded iron cannon. JOM 56(1):17–23. https://doi.org/10.1007/s11837-004-0265-5

Blakely TA (1861) Rifled Ordnance. Roy United Serv Inst J 4(15):397–423. https://doi.org/10.1080/03071846109416731

Bull GV, Murphy CH (1988) Paris Kanonen–the Paris guns (Wilhelmgeschütze) and project HARP: the application of major calibre guns to atmospheric and space research. E.S. Mittler, Herford Germany

Cleator PE (1967) Weapons of War. Robert Hale, London, UK

Davies J, Shumate J, Hook A, Walsh S (2019) The medieval cannon 1326–1494. Bloomsbury Publishing, Oxford, UK, New Vanguard

Farrar CL, Leeming DW (1982) Military ballistics—a basic manual. Battlefield weapons systems and technology series. Brassey's Publishers Limited, Oxford, UK

Finlayson WH (1948) Mons Meg. Scott Hist Rev 27(104):124–126

Fisher J (2020) History of the word howitzer. http://howitzerhistory.weebly.com/etymology.html. Accessed 5 July 2020

Haskell JR (1881) Accelerating Gun. US Patent

Haskell JR (1892) Multicharge Gun. US Patent

Johnson W (1991) Some monster guns and unconventional variations. Int J Impact Eng 11(3):401–439. https://doi.org/10.1016/0734-743X(91)90046-I

King JW (1877) Report of Chief Engineer J. W. King, United States Navy, on European ships of war and their armament, naval administration and economy, marine constructions and appliances, dockyards, etc. etc. [United States] 44th Cong., 2d sess. Senate. Ex. doc. 27, vol 273 p. Govt. Print. Off., Washington

Lewtas I, McAlister R, Wallis A, Woodley C, Cullis I (2016) The ballistic performance of the bombard Mons Meg. Defence Technol 12(2):59–68. https://doi.org/10.1016/j.dt.2015.12.001

Magrath P (2015) Ordnance, B. L., 18-inch Howitzer, Mark I: The last of the super-heavies. Arms Armour 12(2):181–195. https://doi.org/10.1080/17416124.2015.1130900

Mallet R (1856) On the physical conditions involved in the construction of artillery, and on some hitherto unexplained causes of the destruction of cannon in service. The Transactions of the Royal Irish Academy vol 23, pp 141–436

Murphy CH, Bull GV (1966) A review of project HARP. Annals of the New York Academy of Sciences, vol 140. https://doi.org/10.1111/j.1749-6632.1966.tb50970.x

Murphy CH, Bull GV, Boyer ED (1969) HARP vehicles as meteorological and geophysical probes. Annals of the New York Academy of Sciences, vol 163. https://doi.org/10.1111/j.1749-6632.1969.tb13060.x

Murphy CH, Bull GV, Boyer ED (1972) Gun-Launched sounding rockets and projectiles. Annals of the New York Academy of Sciences, vol 187. https://doi.org/10.1111/j.1749-6632.1972.tb48337.x

Newton I (1731) A treatise of the system of the World: by Sir Isaac Newton. Translated Into English. F. Fayram

Parkes O (1990) British battleships, "Warrior" 1860 to "Vanguard" 1950: a history of design, construction, and armament. Naval Institute Press

Romanych M, Rupp M (2013) 42 cm "Big Bertha" and German siege artillery of World War I. New Vanguard. Osprey Publishing Ltd, Oxford, UK

Smith R (1985) HM Tower Armouries: wrought iron cannon project. Historical metallurgy 19(2):193–195

Sullivan BR (1988) A fleet in being: the rise and fall of Italian sea power, 1861–1943. Int Hist Rev 10(1):106–124

Zaloga SJ, Dennis P (2016) Railway guns of World War II. New Vanguard. Osprey Publishing Ltd, Oxford, UK

# Chapter 5
# Making a Gun

One of the more recent privileges I have enjoyed as part of my day-to-day work is visiting a site in Australia where gun manufacture has occurred for over 100 years. That site is in Lithgow, New South Wales, which in its heyday employed 6000 people. The factory is a shadow of its former self, which was once the heart of the town of Lithgow. However, it still manufactures guns, and in fact the most recent addition to the Australian Defence Force's arsenal: the Steyr F90 or 'Lithgow F90'. The factory is positioned in the remote but picturesque blue-mountain region simply because back in 1912, it was impossible to fly bomber aircraft over the mountains to attack the military manufacturing depot. There has been astounding changes in the aerospace industry which would make that assumption ludicrous in this day and age and I would like to say that there have also been leaps and bounds in the technology of guns—but there simply has not. There have been some developments in the manufacturing of firearms, and we will look at these in this chapter.

## 5.1 Bespoke Home-Made Shooters

3D printed guns have been of large concern in recent years with the worry of rapid proliferation and untraceable weapons getting into the wrong hands. However, many guns can be manufactured at home with the right know-how, and the correct equipment such as a lathe or milling machine. These machines are old technology compared to additive manufacturing machines but they are versatile. Cheap, mass-produced single-shot firearms even date back to the Second World War when the Liberator was manufactured to arm the resistance fighters in France. This firearm was crude—but did its job.

Efficient homemade devices have generally been reserved for fictional writers and filmmakers. In the 1993 film 'In the Line of Fire', John Malkovich played the role of an ex-CIA operative by the name of Mitch Leary who was intent on assassinating the

P. J. Hazell, *The Story of the Gun*, Springer Praxis Books,
https://doi.org/10.1007/978-3-030-73652-1_5

President of the United States. In the film, Malkovich's character makes a double-barrelled polymer composite pistol that would not be detected in a metal detector, and would allow him to sneak into a presidential gathering and expedite assassination. In fact, by all accounts the pistol was so realistic that it was destroyed after filming had ceased, probably due to the film-makers' concerns that they were somehow breaking local firearms legislation.

Another famous homemade 'Hollywood gun' was the silenced shotgun used in 'No Country for Old Men', directed by Joel and Ethan Coen and starring Tommy Lee Jones, Javier Bardem and Josh Brolin. In the film, a 12-gauge Remington Model 11–87 Semi-Auto shotgun was used by Bardem with a silencer. However, no such weapon exists with a silencer; the assumption from the movie is that the silencer was made by the psychopathic assassin (Anton Chigurh) in his garage. Of course, any system that propels an object can be used as a weapon. In rather gruesome scenes, Chigurh also uses a 'captive bolt gun' which is pneumatically operated, to attack his victims. With captive bolt guns, compressed air, or a blank cartridge, are used to propel a spring-loaded bolt at the victim's head. The bolt never leaves the gun and is retracted after the firing ready for the next victim (hence the bolt is 'captive'). Captive bolt guns are intended to render animals unconscious and are commonly used in various forms in the abattoir industry.

## 5.2  Mass Production

Guns are big business. By some estimates, there are over 1 billion guns in circulation in the world today and most are owned by civilians. They need to be reliable and safe. Early guns were prone to jamming whether due to poor manufacturing standards or poor-quality materials. Dimensions and the clearance between moving parts are especially important. The adoption of limits, fits and tolerances by the industry enabled the mass production of weapons. The tighter the tolerances, and designs to take into account variability associated with mass production, the more reliable weapon systems became.

The principal material for the barrel and other working parts of a gun is steel. Sure, we have seen previously there have been some wooden, bronze and brass attempts and even more recently a plastic offering in terms of the 3D-printed gun. Peripheral components have been made from other materials. For example, we have seen that the M 777 155-mm Howitzer employs a titanium alloy structure. The 1953 Garrington gun employed a 1-inch thick glass epoxy protective shield, for example. However, steel is a good all-rounder in terms of its properties. That is, it is a material that is strong, tough and we have learned how to tweak the recipe and the processing route to achieve better strength and toughness. It is also cheap.

## 5.2.1  A Brief History of Steel

As we have seen, some of the early siege cannons or bombards were made from iron or bronze (a copper-tin alloy in the proportions of 88% copper and 12% tin). Bronze was principally used as it has a lower melting temperature than iron making it easier to cast into shape. It is also corrosion resistant whereas iron is prone to rust. Iron is also brittle, whereas early bronze guns would show signs of strain before there was a catastrophic failure. Eventually, however, bronze became less popular in favour of iron. Compared to iron, bronze is weak—having about a third of the strength. Furthermore, the constituents of bronze became more difficult and expensive to source as tin supplies became disrupted due to the long trade routes required to carry tin.

Originally iron was regarded widely as, quite literally, "a gift from the gods". Some of the earliest finds of artefacts were made from meteoric iron. These were discovered in the regions of Iran, Iraq and North Africa. Nevertheless, it is highly unlikely that these people were able to connect the arrival of a shooting star with that of iron. The most famous find was that of a dagger which was made from a meteorite comprising of an iron-nickel alloy which was found in the tomb of the Egyptian pharaoh Tutankhamun who lived around the fourteenth century BC (Carter and Mace 1923). Meteoric iron was used widely up to the beginning of the second Millennium BC. It was worth up to 7 or 8 times that of gold at that time. If nobility were hosting a dinner party, only the riff-raff would get the gold cutlery, whereas the VIPs would get the iron cutlery! How times have changed.

In ancient times, iron would have been produced by smelting from the natural iron ore in a type of furnace called a bloomery and hence the product (which was a spongey-looking mass of porous iron) was known as bloom iron. This would have been heated to a 'white heat' (1400–1450 °C) and worked to produce iron ingots from which the knives, swords and shields were produced. The hot forging process was a laborious process and was repeated many times to produce products of suitable quality. The result was sometimes called a 'soft bloom'.

Steel was much better for weaponry. Steel was thought to originate from within the Hittite state. Not many people realise the extent of the Hittite influence, which seem to have been mostly lost to history. They were once an impressive people, centred on the mountainous region in Anatolia and spreading out at the height of their strength through a large proportion of modern-day Turkey and to the East through Syria. Most of what was known of the Hittites prior to the archaeological work done over the last 150 years was to Biblical references as well as minor records of their contact passed down by other peoples. They were a tribe that were well-known for their strong swords and shields and classed as a warrior tribe. Of all the most famous was probably Uriah the Hittite who was regarded as one of King David's mighty men in the Old Testament (2 Samuel 23v39), and he may have derived this title from the quality of his weaponry. However, it is difficult to believe that they had the ability to control the carbon content to such a degree to produce steel (less than 2.1% carbon).

Remarkably, even small changes in the percentage carbon content can substantially affect the properties of the steel.

The English word for steel is derived from the old English adjective 'stylen' in c. 1200, that is describing a form of metal that is strong. The modern form of the word is mentioned four times in the Old Testament of the King James Version of the Bible (compiled in 1611)—possibly in relation to a 'strong metal'. As the prophet Jeremiah noted, 'Shall iron break the northern iron and the steel?' (Jeremiah 15v12). However, by the time we get to the 20th C, it was largely understood that 'steel' was essentially an alloy of iron and carbon.

There is evidence that the process for making high carbon steels using crucible technology originated in India. This steel was known as wootz steel. The first evidence of quality control was also with wootz steels. Archaeological digs have revealed excess ingots that appeared to have been discarded due to their excessive carbon content and hence brittleness. Carbon is important for strength but too much of it compromises toughness and makes the material brittle. Wootz steel finds appeared to have carbon contents in the region of 1.7–2.0%, which just about classifies it as a steel (Gnesin 2016). Wootz steel became the source for the famous Damascus steel that took strength and toughness to a whole new level. Damascus swords became famous for their hardness and their ability to retain their cutting edge as well as the characteristic pattern that was left on the surface of the blade. Remarkably, Damascus steel and Damascus blades had a carbon content of between 1 and 2%. Steels of this carbon content are hard but very brittle. The question is, how was the brittleness removed? It has been since discovered that by a complex process of mechanical working at suitable temperatures, the microstructure was converted to a fine grained structure which behaved in what we call a 'super-plastic' fashion. That is, the material is able to be squeezed and squashed to a much higher degree than would have otherwise been possible. Furthermore, the other remarkable feature of Damascus steel is that through processing, carbide nanowires of approximately 10–20 nm in diameter and several hundred nanometres in length are formed. Of course, in the middle medieval times nanotechnology would have been 500 years off from being discovered. Yet, the blacksmiths were able to skilfully employ the benefits of nanotechnology in this steel. In fact, during this time, the actual role of carbon in the manufacture of steels would have been a complete mystery.

For the steel used in early guns, the percentage of the carbon was restricted to 2.5% (which is actually cast iron). Modern steels comprise of carbon compositions of no more that 2.1%. Towards the end of the 19th C., the 'gun community' were not too sure about the exact nature of the steels to which they were subjecting enormous pressures. There existed vast differences of opinion on the role of the carbon within the iron (Bruff 1896). Harder steels were referred to as having 'dissolved' carbon whereas, softer steels had 'undissolved' carbon. Alternatively, it was described as 'fixed' carbon for hard steels and 'free' carbon for the soft steels. However, it was thought that the carbon would transform from one form to another through heat treatment and cooling.

## 5.2.2   The Nature of Steel

First, a little bit of materials science. Steels are what we call polycrystalline materials. That means that the atoms are arranged in an ordered fashion to form grains or 'crystals'. Multiple crystals link together to form a polycrystal (poly means 'many'). Many metals are like this. The other form of material commonly encountered are those materials that are amorphous. With these materials, the atoms are not arranged in any particular order and there appears to be chaos at the atomic level with the atoms distributed all over the place. Polymer materials are commonly like this.

One of the foremost engineers of the time who developed the terms 'fixed' and 'free' carbon was a Swedish engineer J. A. Brinell (1849–1925). Brinell is best known today for the Brinell hardness test, which he proposed in 1900. However, he was also a shrewd metallurgist.

To determine the effects of heating and cooling on the change of properties, Brinell made a series of experiments. Using a steel that contained about the same percentage of carbon as gun-steel, he heated bars of the steel to different temperatures and cooled them at different rates. The bars were subsequently broken and the fracture examined. He found that there were two states of the carbon one which he called 'free carbon' and which was associated with soft steel, and the other, 'fixed carbon', was associated with hard steel (Lissak 1907).

Brinell noticed that for each steel that he called 'hard' and 'soft', there was a certain temperature, called the critical temperature, to which if the steel was heated, and suddenly cooled, resulted in a change in properties. Brinell thought that all the carbon was consequently 'fixed', and the surrounding structure became amorphous. We now know that the structure would not be amorphous and Brinell's 'critical temperature' was referring to the 'austenitic temperature'—a temperature that induces a change in the steel. He also thought that with 'hard steel', if heated to this critical temperature and cooled slowly, the steel would acquire a 'crystalline structure', and all the carbon would become 'free'. He was partly correct here as we will soon see.

So, what were these 'free' and 'fixed' carbon contributions? We actually now know that unlike many non-ferrous alloys, the cooling rate plays a large part in how the microstructure (and the resultant mechanical property) is changed. Heating a steel up to 900 °C and then cooling it quickly will have a different effect than cooling it slowly—depending on the composition of carbon. And, this is all to do with the special property of iron called 'polymorphism'. This unusual property has led to steel's convoluted history. Polymorphism is the ability of a material to exist in more than one crystal structure and for iron (Fe), this occurs when the temperature is increased. That is, the relative positions of the atoms rearrange themselves to form different crystal structures as the temperature is changed. Increasing the temperature results in different forms of iron that are denoted by Greek letters $\alpha$, $\beta$, $\gamma$, $\delta$ (early metallurgists simply loved Greek, it seems!). That is, except that we now know that $\beta$-iron does not exist. Early investigators observed that the magnetic behaviour of iron disappeared at 768 °C and attributed that observation to a phase transformation. They assigned this transformation the '$\beta$-label'. Later it was discovered that the loss of the magnetism was due to other factors. However, by that time, the other forms of iron were well and truly established in the lexicon. At room temperature, the stable form of iron is called ferrite, or α-iron (or α-ferrite) and the atoms are arranged as a body-centred cubic (BCC) [1] crystal structure. At 912 °C, ferrite undergoes a polymorphic phase transformation to a face-centred cubic (FCC) γ-iron (think of the atoms arranged like on a dice where each face is a '5'). Increasing the temperature to 1394 °C results in the reversion back to a BCC structure with the formation of δ-ferrite. This finally melts at 1538 °C. The δ-iron is virtually identical to α-iron except over which the temperature range it exists.

The rate of cooling from elevated temperatures has a large effect on the structure of the steel, which in turn affects the properties. For example, slowly cooling a plain carbon steel from around 850 °C will result in something called pearlite being formed. This was so-called as the structure has the appearance of mother-of-pearl under a microscope at low magnifications. Pearlite is an alternate plate-like mixture of α-iron and iron carbide ($Fe_3C$). The iron (α) phase is quite soft (relatively speaking, I should say) with a local hardness of about 90 VHN (Vickers Hardness Number[2]) whereas the bulk pearlite is harder with a typical hardness of 250 VHN. Slow cooling is done in air, a process known as normalizing. Rapid cooling or 'quenching' is done in water or oil. Rapidly cooling the same plain carbon steel in water will result in another phase known as martensite which has a unique crystal structure BCT (Body Centred Tetragonal). This crystal structure is highly strained that lends itself to brittle behaviour. Martensite is named after the German engineer Adolf Marten (1850–1914) who worked in the fields of metallography and mechanical testing. This is the hardest constituent obtained for a given steel. Martensitic microstructures consist of fine needle-like structures. This results in a steel that is hard but brittle and therefore they need tempering, which is a process of reheating and slow cooling, to allow for an improvement in ductility. Another constituent that can be formed on cooling in alloy steels is bainite. Through careful choice of alloying elements and tempering temperatures it is possible to produce a super-strong bainite steel with nano-sized grain sizes. This makes the steel very strong with tensile strengths of 2 GPa quite possible.

### 5.2.3 Carbon Limits

It is the percentage by weight of carbon that defines whether a material is a steel, a cast iron or even an iron carbide. Steel is defined by an understanding of the iron-iron carbide phase diagram. Work started on this phase diagram around 1870 but an initial

---

[1] That is, atoms of iron are arranged at the four corners of a cube and one atom is located in the centre.

[2] VHN = Vickers Hardness Number which is a scale based on the imprint of a diamond indenter into the material. The indenter is loaded with a pre-defined mass and the larger the indentation the lower the VHN and the softer the material.

version was not fully completed until 1920 and even this had some errors. Further-more, even in the late 1940s there was still some problems with our understanding of the phase diagram which still persist and find their way into textbooks! (Wadsworth 2002).

In some ways, it is because of the nature of iron, and how it behaves under increasing temperature, that mankind has been able to develop new and improved technologies with steel. Iron's polymorphism underpinned the industrial revolution. As we have seen, it is now well-known that heat followed by cooling, either rapidly or slowly, affects the microstructure of steel and this is due to the fact that iron undergoes physical microstructural changes as the temperature is changed. Further, as the microstructural changes occur in the steel so do the mechanical and physical properties change. And these properties can be further tweaked by working the material (or indeed, further heat treatments). This latter fact has been well known to blacksmiths for centuries.

## 5.3  Coping with the Pressure

There is no doubt about the fact that guns generate huge amounts of pressure in their chambers. These pressures can reach up to 600 MPa for a tank gun. In the early days of gun manufacture the only way that pressure could be contained was by using thick sections of material. As a gun fires, the internal pressure will attempt to cause the barrel to bulge and the barrel wall material is subjected to tension around the circumference. This stress is referred to as the 'hoop stress'. As we have seen, it turns out (according to Lamé) that increasing the wall thickness of the gun leads to a reduction in the overall maximum hoop stress seen by the steel. That is the circumferential stresses that act around the gun wall if you looked at the gun end-on. And, if you know a "Hooper" you can almost certainly guess that their ancestors were involved in some form of wooden barrel manufacture and produced or installed the iron hoops that kept the wooden 'coops' together. They were probably made by the "Coopers".

To unpack this somewhat further, Fig. 5.1 shows a schematic of a gun barrel with the option to increase the outer radius from $r_1$ to $r_2$. A fixed pressure is applied to the inside of the gun barrel. On the right of this figure the hoop stress varies in the wall thickness as the radius is increased. You will notice that for a given pressure, increasing the outer radius of the gun barrel reduces the overall tensile hoop stress in the material. So, this is why very early guns had to be enormous with some of the muzzle loaded cannons having very large chamber and barrel thickness values to accommodate the pressure.

Most small high-velocity small arms weapons would see a pressure of around 380 MPa and it is therefore important to get the design correct to ensure that the pressure is contained. After all, the chamber is very close to the shooter's head! Most gun designers calculate the required wall thickness based on a maximum proof pressure or even design pressure (which should be higher). This allows for a factor of

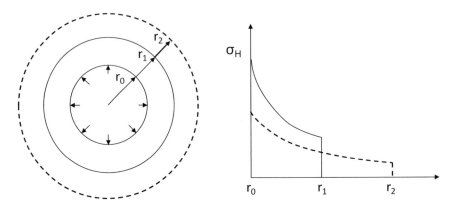

**Fig. 5.1** Schematic of gun barrel with a single calibre ($r_0$) and two different outer radii (left) and a graph showing the variation of stress in the barrel ($\sigma_H$) for the two cases. Increasing the wall thickness from $r_1$ to $r_2$ leads to a reduction in the hoop stress as shown by the dotted line in the graph. *Source* Author

safety when it come to mean operating pressures. And, only when the gun is misused can catastrophic events occur, and even then, the gun should be designed to protect the shooter from a disastrous outcome.

For large guns, a gun tube was historically over-wrapped with wire. Long lengths of high-tensile wire would have been wound onto a tube under constant tension. The tube had the necessary rifling and once the wire had been wrapped, the gun would be jacketed with multiple hoops of steel that would enclose the wire windings (L'H Ruggles 1916). The high tensile wire would provide a compressive pre-stress to the inner tube of steel which would go some way of reducing the tensile stresses seen by the gun during firing. HMS Furious, which had 18-inch guns, used this technique. Even then, the barrel's wall thickness was quite enormous. Examples of guns that were made by this technique are provided by the Woodbridge and Crozier guns (see Fig. 5.2). The Woodbridge gun consisted of an inner tube, *t*, wrapped with wire as shown (*w*). Over the rear part of the tube is a jacket, j, made of longitudinal steel bars

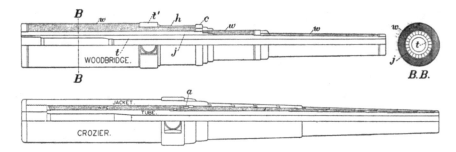

**Fig. 5.2** Woodbridge and Crozier guns that both utilised the wire-winding approach to preloading the tube. Reproduced from (Bruff 1896)

**Fig. 5.3** A composite barrel comprising of a fluted steel inner tube for cooling and an outer part that acts to compress the inner tube

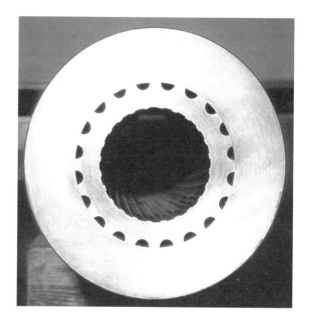

of wedge-shaped cross-section. This jacket was wrapped with the wire $w$, under such tension as to strongly compress the inner tube. With the Crozier gun (lower figure) the tube was compressed initially beyond its elastic limit by the wire. The wire on the chase of the gun was covered by thin hoops with the purpose of protecting the wire (Bruff 1896).

Another approach was to shrink-fit an outer tube onto an inner tube. The effect would be similar to the wire-wound approach providing a compressive pre-stress to an inner tube of steel. This also provided an opportunity to cool the barrel with flutes that run down the length of the tubes. An example is shown below in Fig. 5.3.

However, by far the most common way of minimising the hoop stress during firing is by a process of autofrettage. Autofrettaging makes use of the fact that if you apply an extreme load at the inner surface of a metal, it plastically deforms. Therefore, the inner part of the gun barrel is plastically worked whereas the outer portion remains elastic. This is quite normal in the manufacture of large guns but is perhaps somewhat disturbing to many engineers who have been educated to believe that all structures should operate within their elastic region. You see, when you plastically work a metal, you are in fact causing damage to the structure of that metal. Nevertheless, it serves an end-purpose in that during the process, the whole tube is being stretch elastically, whilst the inner part is stretched plastically. The outer portion of the steel will want to attempt to return back to its original starting point once the load has been removed. Thus, the plastically worked section is subjected to compressive loading. Therefore, when the gun fires the steel will try and stretch. The pre-load reduces the amount of tension the barrel ultimately sees. Autofrettaging can also lead to a weight reduction in the barrel and as much as 60% weight savings are possible.

There are two methods that are often used to achieve an autofrettage in a gun: These include: swage autofrettage and hydraulic autofrettage. The latter is achieved by placing the gun barrel over a mandrel and applying a large internal pressure to the internal surfaces of the barrel. This leads to the desired plastic deformation of the internal surface. Enormous pressures are required to achieve the desired effect (in excess of a 1 GPa). And, consequently, very large pipe work with enormous wall thickness is required to provide the hydraulic fluid to the desired location. This is currently very common with European manufactured gun barrels of the larger calibre variety. With a swage autofrettage approach a tungsten carbide swage is pulled through the gun barrel and contact surface of the swage is angled to achieve the desired plastic flow of the internal surfaces.

The amount of pre-load that is put into the steel is illustrated by Alistair Doig (Doig 2002). He reported an occasion where he was making a longitudinal cut in a half-metre length of a 120 mm tank gun barrel from near the breach for a metallurgical investigation. He reported that when the longitudinal cut was nearly finished, the barrel section sprang open into a 'C' shape with a massive bang. Accordingly, the area of the fracture was about 600 mm$^2$ and this equated to a force of about 100 tons!

## 5.4  Making a Gun Barrel

Large gun barrels used in tanks guns, for example, start life as an ingot. The ingot is reduced in mass from around 7 tonnes to around 4.5 tonnes by the removal of unsound porous material. After which, it is ready for 'hot hollow forging'. This involves the ingot being pierced to form a hollow preform through which a solid steel cylinder called a mandrel is fitted. At this stage, the temperature of the preform is around 1200°C. The preform is then systematically fed into a large industrial press that shapes and squeezes the forging into the shape of the barrel over the mandrel. The steel needs to be kept at an elevated temperature during this process and therefore there are intermediate furnace reheats that are required. The forging is then slowly cooled to 300°C to allow for outgassing any hydrogen. Hydrogen can be deadly to gun-steels. Atomic hydrogen (as opposed to molecular hydrogen, $H_2$) is known to diffuse into the structure of the steel, literally dissolving into the steel, and thereby reducing its ability to plastically deform. Ultimately, during this process the steel becomes more brittle and therefore removal of hydrogen is essential. The steel is then given a rough machining and then subjected to another round of heat treatment to optimise the strength and ductility.

The principle of making a small-arms rifle is selecting the correct steel and machining it to the correct length. The steel is drilled oversize and then the steel is hammer-forged to produce a cylinder that is approximately 1.5 times the length of the original. This process is rapid. It is commonly accomplished by 4-6 specifi-cally shaped tungsten hammers that literally bang the barrel into shape. A lubricated high-strength mandrel is placed inside the barrel and dragged rearward to coincide with the position of the hammering and to form the internal shape, that can include rifling. The hammering process occurs in around about a minute after which further machining and dressing will occur.

## 5.5 Avoiding Gun Wear

When a gun fires there are very high temperatures that are realised in the bore. The temperatures of the working surfaces can reach as high as 900 °C whereas the gas temperature of the propellent can reach as high as 2200 °C. At around 700 °C we find that the microstructure of the steel begins to change. And in fact, we see that the tempered martensite on the very surface will transform to austenite. This leads to a contraction in volume of about 4%. Then on cooling between firings this austenite will transform back to martensite giving a roundabout 4% expansion in volume. These very localised volumetric changes and reversals lead to craze cracking. This is an inevitable result of using steel for the gun barrel and happens regardless of the actual grade that is chosen (Doig 2002).

One of the ways to protect the steel from the high temperatures is by chrome-plating the bore. Chromium has a very similar coefficient of thermal expansion to steel and it is also able to cope with the high temperatures of propellant ignition and burn quite nicely. Matching the coefficient of thermal expansion of the steel is most important as any major differences in this value will curtail the life of the gun as the plating will attempt to separate during the repeated thermal cycles due to firing. If the rifle is chrome-plated then that will an electro-plating process where the chromium is deposited on the inside barrel after rifling. This will produce a layer between 8-15 microns in thickness and extends the life of the barrel. Chrome-plating can be a nasty business and, in some processes, uses hexavalent chromium. This is the same material that was the subject of litigation case involving Erin Brockovich that was made into a blockbuster movie of the same name starring Julia Roberts. The case (Anderson, et al. v. Pacific Gas and Electric, file BCV 00300) alleged contamination of drinking water with hexavalent chromium (also written as "chromium VI", "Cr-VI" or "Cr-6") in the southern California town of Hinkley. Gun manufacturers therefore need to adhere to strict health and safety rules when applying this product and this ultimately has an impact on the price of the finished barrel. Sure, chrome-plating can extend the life of the gun barrel but can lead to doubling or even tripling of the price.

## 5.6 The Old, the Odd, and the Outstanding

There have been several interesting firearms over the years, and it is prudent to finish this Chapter by looking at several of these. There have been some arguably unusual guns that we have already touched on such as Armstrong's 100-tonne muzzle-loader, Gerald Bull's 'Supergun' and Hitler's V3. There are also the two-stage guns and the three-stage guns. There have been several guns that deserve their place in the history books as the stand-out guns. There are probably too many to mention; however here are a few that have piqued my interest.

There is nothing more that can indicate the success of a weapon than its longevity in service. One of the longest serving pre-modern rifles that has been used was probably the British 0.75-inch-calibre Long and Short Land Pattern Musket, popularly known as "Brown Bess". This weapon was found on the front line of the British Military campaigns from 1722-1836. Subsequent to that, upgraded versions were used in the Crimean campaign (1853-1856) and the Indian Mutiny (1857-1858). It was also modified and used by the Confederacy in the American Civil War (1861–1865). Why this weapon was called Brown Bess nobody actually knows. And in fact, it did not get its nickname until the 1700s. There are several theories, however. The 'Brown' may refer to the Brown walnut stock or the brown varnish applied to protect the metal and wooden parts. 'Bess' may well have referred to Queen Elizabeth I who was known as 'Good Queen Bess' and of course she was known to be a hard and demanding woman, but she died over 100 years before the musket's adoption. There are other theories too, but most are based on conjecture.

It is also worth mentioning the British 0.303 Lee Enfield No. 1 Mk III rifles which was the British Army's standard rifle from its official adoption in 1895 until 1957 and is still in service in India and Nepal today. Indeed, many an old soldier will still lament its demise. Its successor, the Lee Enfield No. 4 Mk 1 was equally as successful and adopted in 1931 and apparently still found in operation with reserve forces in many African nations as well as with the Canadian Rangers.

In the US, there are have been a couple of long-serving guns that have stood the test of time. None, more so than the M1 Garand. This was a 0.30-calibre (7.62 mm) weapon and was adopted in 1936 by the US Army and was used in two Wars. Additionally, the 0.30-calibre M1903 Springfield which was issued in 1906 while still being used during the Korean conflict until at least 1954. Additionally, the 7.62 mm M14 which was standardised in 1957 and 1st issued in 1959, is still in use today. It is even being presently carried by US military Academy West Point cadets as a ceremonial weapon.

The German 1898 Mauser, which has been licenced built in numerous variants by many countries, is still in use today as a sporting rifle. Some regard this as the ultimate bolt-action rifle and it is estimated that over 100 million models have been made— similar to the AK 47. The Soviet 7.62 mm AK 47 assault rifle or 'Kalashnikov'—so named after the inventor Mikhail Timofeyevich Kalashnikov (1919–2013) is another famous and long serving weapon and arguably more prolific than the 1898 Mauser. This was adopted in 1949 but not widely available to the Soviet army until 1956. It is of course the weapon that has been widely used by many insurgents and terrorists over recent years and famous for its simplicity and robustness. It is estimated that over 100 million AK-47 assault rifles have been manufactured and this weapon has probably killed more people than any other weapon. It is resilient and very rarely jams. It is so light and the reçoil so slight and efficient that even a child can fire it with ease. Sadly, this means that children have been recruited to be deadly killers with this weapon in numerous conflicts.

And then there are the unusual guns. One of the most unusual guns was made specifically for Louis XIII. Unusually it had a barrel that was shaped as a fleur-de-lys of France and thereby a flattery to Louis. Nice! However, the trouble was that one would not expect a projectile to travel well or indeed penetrate its target efficiently given the odd cross-section. Very clearly a smooth-bore weapon (fleur-de-lys shaped projectiles cannot spin) and it was highly likely that the projectile would have got stuck during firing. Barrels have not always been circular. In fact, an oval-bore cannon was made in Germany around 1625. There are also some relics in the Tower of London that suggest that certain mortars had muzzles shaped like a letterbox that fired bars of iron.

Other oddities include the use of materials other than steel to construct a gun. It is well known that wood was used in China as the main barrel material. The Paris Artillery Museum houses a wooden gun from Cochin in China. Two halves were hollowed out of wood and banded together with iron strips (Carman 2004). Clearly this is a relatively cheap way to make a cannon. However, it is very likely that after only a few shots the barrel would be prone to failure, or at the very least gross erosion. Famously, Henry VIII used wooden artillery at the siege of Boulogne in 1544. However, leather was also used. Venetians were thought to use leather in conjunction with hempen rope and Gustavus II Adolphus of Sweden (1594–1632) used leather cannons extensively. These comprised of thin tubes of copper that were encased in leather hide. The principal advantage of these leather guns was the mass was a mere 40 kg, meaning that they could be handled by two men, unlike the cannons and demi-cannons of the time (Cleator 1967). A leather cannon was fired three times at Kings Park, Edinburgh in October 1788. Even paper has been used. It was written that in 1640 an artificer of Bromsgrove near Worcester was very successful in making firearms of paper and leather (Greener 1910). Apparently, these were recommended because of their lightness and strength. It is highly unlikely that these guns would have lasted long, however. Probably the oddest materials that has been reported to have been used as a gun barrel material, is ice. Ice was apparently used in 1740 to construct a mortar in St Petersburg, which was repeatedly fired (Johnson 1991; Ffoulkes 1969). It could not have lasted long, however. Ice was also a 'material of interest' during World War II with the development of Pykrete, a solidified mixture of wood shavings and ice. Further, Pykrete, which was named after its inventor, Geoffrey Pyke (1893–1948), had structural significance to the extent that it developed a reputation for protection against small-arms fire. Pyke even had ambitions to build a massive floating aircraft carrier out of the material. That project was called Project Habakkuk and had the blessings of Mountbatten and Churchill. However, despite early enthusiasm, the project sunk (literally) and never came to fruition due to refrigeration costs.

The smallest gun in the world has to go to the 2.7 mm Kolibri (German for Hummingbird). See Fig. 5.4. This was the ultimate in self-defence weapons and easily concealed weighing in no more than 200 g when fully loaded and with an overall length of 7 cm. Patented in 1910 by an Austrian watchmaker called Franz

**Fig. 5.4** The 2.7 mm Kolibri—this specimen was photographed by the author at the Lithgow Small Arms Factory Museum, Australia. *Source* Author

Pfannl and introduced in 1914 and was marketed as a lady's self-defence weapons that was compact enough to fit into a handbag. The muzzle energy was 4.0 J and therefore unlikely to be able incapacitate a persistent mugger and probably unable to penetrate the thick clothing. Nevertheless, if aimed at the eye it was dangerous—although not very accurate given that it was a smooth-bored pistol.

# References

Bruff LL (1896) A text-book of Ordnance and gunnery: prepared for the use of cadets of the U.S. Military Academy. Wiley, New York

Carman WY (2004) A history of firearms: from earliest times to 1914. Dover Publications

Carter H, Mace AC (1923) The tomb of Tut-ankh-amen: discovered by the late Earl of Carnarvon and Howard Carter, vol 1. Cambridge Library Collection—Egyptology. Cambridge University Press, Cambridge. 10.1017/CBO9780511722349

Cleator PE (1967) Weapons of war. Robert Hale, London, UK

Doig A (2002) Military metallurgy. Maney Publishing, London, UK

Ffoulkes C (1969) The gun-founders of England: with a list of English and Continental gun-founders from the XIV to the XIX Centuries. Arms and Armour Press, London, UK

Gnesin GG (2016) Iron age: origin and evolution of ferrous metallurgy. Powder Metall Met Ceram 55(1):114–123. https://doi.org/10.1007/s11106-016-9786-z

Greener WW (1910) The Gun and its development, 9th edn. Bonanza Books, New York, USA

Johnson W (1991) Some monster guns and unconventional variations. Int J Impact Eng 11(3):401–439. https://doi.org/10.1016/0734-743X(91)90046-I

L'H Ruggles C (1916) Stresses in wire-wrapped guns and in gun carriages. Wiley, New York

Lissak OM (1907) Ordnance and Gunnery: a text book, 1st edn. Wiley, New York
Wadsworth J (2002) Ancient and modern steels and laminated composites containing steels. MRS
    Bull 27(12):980–987. https://doi.org/10.1557/mrs2002.305

# Chapter 6
# Propellants and Explosives

To propel a projectile along a gun barrel, the propelling medium should be able to do three key things: Firstly, the propellant is required to release its energy in a controlled fashion so that the pressure is containable (that is, that the gun is not damaged) and that pressure is maintained on the base of the projectile. Secondly the propelling material should be completely consumed leaving little-to-no residue that can foul the weapon and, thirdly, the propelling medium should be able to be reliably initiated.

We have seen that for guns to function it is necessary to use some form of chemical energy to propel the projectile along the barrel and this initially was black powder. This broke down in a violent exothermic (= heat generating) reaction to produce gases which in turn could be used to drive (or 'propel') the projectile. This has since been replaced by synthetic nitrocellulose-based compounds. So, what is the science and the story behind these materials?

## 6.1   Why Do Propellants Function?

Gun propellants are simply combustible materials that burn rapidly without external oxygen and convert their current state (solid or liquid) into a gas through an exothermic reaction. The main purpose of a propellant is to burn fast and produce rapidly expanding gasses which act on the base of the projectile. The action on the base of the projectile causes the projectile to accelerate. Black powder, sometimes referred to as gunpowder, was an early form of propellant. These are sometimes referred to as 'low explosives' which burn as opposed to 'high explosives' which detonate. More on that later.

All gun propellants contain nitrogen. In fact, nitrogen is a critical constituent of high explosives too such as TNT, Nitroglycerine, PETN[1] and so on. They also contain

---

[1] Check out the Glossary for the definition of these.

© Springer Nature Switzerland AG 2021
P. J. Hazell, *The Story of the Gun*, Springer Praxis Books,
https://doi.org/10.1007/978-3-030-73652-1_6

the oxygen required for their combustion. It is the addition of nitrogen that is the key to the rapid release of gas. Nitrogen, $N_2$, that is abundant in the air that we breathe, is a very stable molecule due to the strong bonds that exist between the two nitrogen atoms. However, in its nitrate state (that is, when joined to oxygen) it is quite volatile and when incited will break down rapidly to form nitrogen gas. Military explosives are made through a process of nitration whereby the $NO_2$ molecule is added to various organic compounds. So, to form nitrocellulose there is a process that involves mixing sulphuric and nitric acids with cellulose and results in a product with the chemical formula comprising of multiple $NO_2$ sections. Before we get to this process, we need to separate out the nitrogen molecule, think of separating out two strongly-attracted magnets—it takes energy to do so. Therefore, when the process occurs in reverse and the magnets are allowed to be reassemble and when the atoms of nitrogen recombine to form nitrogen gas, energy is liberated. With the magnets, this will be in the form of kinetic energy and sound as they collide together. Hence the energetic nature of these materials.

An important factor for the function of the propellant is the way that it burns it is assumed that on firing the surface of the propellant granule is ignited simultaneously and the granules burn in parallel layers. For example, if the granule is a cylindrical shape it will retain its cylindrical shape and be consumed from the outer edge to the centre. This process was discovered by a French General by the name of Guillaume Piobert (1793–1871) and so this became known as Piobert's law. It is also strongly supported by experimental evidence. As the granule of propellant burns the reaction that ensues converts the solid into the liberated gas. The resulting gas has a high temperature and expands into the chamber at a rapid rate. A small proportion of the liberated energy is conducted into the granule as the surface burns and this raises the temperature of each successive layer, thereby perpetuating the reaction.

Another important factor that came to be understood is the void ratio. This is the ratio between the volume occupied by the physical propellant and the volume of the chamber in which the propellant is contained. The void ratio controls how quickly a propellant breaks down and dictates the design of cartridges and gun chambers. If this value is close to unity, implying that there is only a small gap into which the propellant can 'burn into', then the local pressure increases rapidly during the process. However, combustion rate is very sensitive to pressure. So as the pressure builds up, the burning rate accelerates.

## 6.2   The Development of Propellant

We have already seen in Chap. 1 that propellant, meaning a substance that could 'propel' a projectile, was discovered by the Chinese with the discovery in Europe attributed to Roger Bacon (1220–1292). However modern developments of propellant have led to the development of nitrocellulose-based propellants. Most firearms use what we call a single-base propellant. That is, where nitrocellulose is solely used with perhaps traces of additives. Nitrocellulose is based on a long chain repeating molecule comprising of carbon, hydrogen, nitrogen and oxygen with the chemical formula of:

$$\left[C_6H_7O_2(ONO_2)_3\right]_n$$

Nitrocellulose is made by treating cellulose with a mixture of acids (sulphuric and nitric). This changes the hydroxyl groups (–OH) in the cellulose to nitro groups (–NO$_3$). Nitrocellulose was also known as gun cotton and the main ingredient of smokeless gunpowder. Nitrocellulose was discovered by Henri Braconnot (1780–1855) of Paris in 1833. He found that when nitric acid was applied to starch, sawdust and cotton wool the composition became incredibly flammable. It was merely a chemical curiosity, however, until a Professor Christian Friedrich Schonbein (1799–1868) working at the University of Basle discovered that by the addition of sulphuric acid and combining with nitric acid (mixed in a proportion of three parts of sulphuric to one of nitric acid) an "explosive cotton" was formed. It was patented by Schonbein in 1846. Nitrocellulose and nitroglycerin were discovered within a decade of each other, but neither found widespread use until the 1860s when methods of stabilizing them were devised (Oxley 2003). In fact, initially, the gun cotton was so rapid in its burn-rate that it was widely ignored for its application in firearms, despite the relatively clean burn. Many feared it would simply burst firearms apart. So, gun powder remained the propellant of choice—well into the 20th C, largely out of ignorance to other innovations. Finally, during the 1880s, a French inventor by the name of Paul Vielle (1854–1934) developed a nitrocellulose-powder which unlike previous powders, created little smoke or residue and was able to burn in a stable fashion. Vielle revealed this synthetic "smokeless" powder, known as Poudre B, to the World in 1886.

There were other obvious uses of gun cotton, however and the material was enthusiastically welcomed by mining companies that could make good use of the high burn rate and resulting explosive energy that was released. Perhaps, the less obvious use of gun cotton was its use in surgery as it was found to make an excellent plaster-like material that could be adhered to a wound (Munn et al. 1857). Kind of like an explosive plaster! Collodion, as it became known, was a mixture of nitrocellulose in ether and alcohol and applied to wounds as a surgical dressing. It also led to the development of celluloid photographic plates and later, celluloid film. Understandably, early celluloid films were infamous for being extremely flammable and therefore great efforts were required by the film industry to handle with care.

The development of a propellant was important for the workings of a gun. Ideally the propellant burns without leaving residue. The build-up of residue leads to fouling and jamming of the moving parts. This was the main problem with black powder which would produce a relatively large quantity of solid residue that would plague early gun developers such as Gatling. The other important feature of a propellant is that the flame temperature was minimised during the violent exothermic reaction. This is not only important for the handling of the weapon but also the longevity of the gun. In fact, the choice of propellant is one of the most importance decisions for self-loaders and manufacturers as it has a direct bearing on the velocity of the projectile and therefore range, the longevity of the barrel, the comfort during firing and the cleanliness of the moving parts.

There are essentially 3 types of propellant available. Firstly, there is the single-base propellant, which comprises of mostly nitrocellulose as well as a chemical called a 'moderant' to control the burn-rate and a small amount of stabiliser and plasticiser (for increased shelf-life). Secondly there are double-based propellants that comprise of nitrocellulose and nitroglycerine in a roughly 50:50 split by composition. Again, a stabiliser is added to inhibit chemical breakdown and therefore extend the shelf life. Roughly 9-10% of the propellant will comprise of this material that itself is composed of compounds such as carbamite or diphenylamine. The advantage of adding nitroglycerine is that it increases the power of the propellant which in turn increases the velocity of the projectile. Thirdly, there are the triple-based propellants. During World War II there was a need to improve the longevity of guns. The temperature of the propellant has a large influence on the erosion of the gun tube. The higher the flame temperature of the propellant, the quicker the gun erodes leading to inaccuracy and reduced performance. The Germans discovered that there was less tube erosion when nitroguanidine was used with diglycoldinitrate. The main purpose of these additives was to extract out the heat from the propellant during the ignition and burn process. Nitroguanidine also had the advantage of quenching muzzle flash although smoke formation was more intensive. It also produced more gas and therefore the reduction of heat during deflagration was compensated by an increased gas volume. These became known as triple-based propellants.

Propellants need to be initiated safely and reliably and are done so using initiating compositions. These are very sensitive compositions that are used in devices such as primer caps. Historically, mercury fulminate was used, and later potassium chlorite was added to increase flash and produce hot particles. In the quest to improve the reliability of initiation, antimony sulphide was added to improve the flash and increase the local temperature received by the propellant. Even though the percussion ignition system is over a century old, no single explosive compound has been found which satisfies all of the criteria for small arms initiation. Modern primer caps use mixtures such as VH2 which are comprised of lead styphnate, tetrazine, barium nitrate calcium silicate, lead peroxide and antimony sulphide (Goad and Halsey 1982).

## 6.3  Caseless Ammunition

As we have seen previously, in tank guns, it has been fairly common practice over the past several decades to move away from expensive brass cartridges and use fully combustible propellant systems or caseless systems. To recap, the main component of brass is of course, copper. Copper is an expensive commodity, and this drove many armed forces to recycle their expended cases. The British L7 rifled gun that was initially employed on the M1 Abrams and tanks such as the Leopard 1 and Challenger 1 was the last major tank gun to use brass cartridge cases. Since then, ammunition has been designed around dispensing the use of brass for a combustible

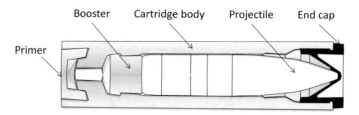

**Fig. 6.1** Schematic sectional view of a caseless cartridge ammunition (calibre = 4.73 mm) for the G11 weapon system ( adapted from (Meyer et al. 2007))

case. Obturation can be achieved by using a complex breech system such as with the British L11 and the L30 tank-gun breeches. Alternatively, a stub case is employed. The name 'stub' implies that the cartridge is much reduced in length and so it is a kind of shortened cartridge case where the length is only 70–100 mm.

There have been several attempts over the years to develop what is called a caseless round or caseless ammunition for small arms applications. In its simplest form this normally means that the bullet is encased in a solid propellant. The advantage is that there is no need for an ejection port to remove the cartridge after firing simply because everything is consumed. The additional advantage lies in the fact that the ammunition is lighter in weight than conventional brass cartridge ammunition. One of the notable examples of this is the Heckler and Koch G 11 assault rifle. With this weapon, which was of a bullpup design[2], the ammunition was 40% smaller in volume than its cartridge counterpart.

If the combustible cartridge case is designed well, the cartridge that surrounds the projectile can add to the ballistic performance of the projectile. So, a high energy material such as nitrocellulose is used as well as some form of structural reinforcement such as kraft-paper pulp (Meyer et al. 2007). The main problem for caseless ammunition design for small arms is overcoming the risk of cook-off at relatively low temperatures due to the nitrocellulose-based cartridge being in direct contact with a hot chamber. There is a risk too that the chamber would get hotter than it would have otherwise done given that brass is a good heat sink and some of the heat energy would be transferred to the cartridge as opposed to the chamber. To get around this with the G 11, Heckler and Koch developed a High Ignition Temperature Propellant (HITP). See Fig. 6.1 for a schematic of the round.

## 6.4   Don't Breathe in!

In turreted systems it is important to avoid excess propellant gasses finding their way into the turret. Post firing, hot propellant gasses would cool and trickle back into the

---

[2]A bullpup firearm is one where the action (that includes the bolt mechanism and extractor port) is located behind the trigger assembly.

turret under the action of gravity and the movement of gasses as the breech door was opened. Propellant gasses are nasty and apart from the risk of hindering sight and breathing, occasionally unburnt propellant can cause a 'flashback' as the breech door is opened. Fume extractors or 'bore evacuators' go some way to solve this problem. They started appearing in the 1950s on tanks such as the Russian T-10 and later on the T-54s and T-55s (Perrett 1987). However, they appear to have been trialled on the American T-34 Heavy Tank in 1945, for which no production orders were placed (Hunnicutt 1988). Intrinsically they are very simple systems. They comprise of a simple cylindrical can that sits around the gun barrel and in which there are four to six angled holes drilled to allow for the passage of propellant gases into the cylinder from the bore. As the projectile travels along the bore, high pressure propellant gases find their way into the fume extractor until the pressure is equalised. After the projectile has just left the gun and the pressure in the bore is reduced, the high pressure in the fume extractor is released back into the bore. The holes through which this is accomplished are angled towards the muzzle and therefore the gas is forced in the direction of the muzzle. Further, as the propellant gas cools the volume of these gases reduces and so a small but measurable reduction in pressure leads to air being dragged from the turret when the breach door is opened. This all helps for maintaining a clean and friendly environment in the turret. An example of the way they work is shown below in Fig. 6.2. The process is shown as follows: (a) high pressure propellent gas is formed behind the projectile, (b) the gases are forced into the fume extractor as the projectile passes by, (c) after the pressure starts to drop, the gases are accelerated from the fume extractor towards the muzzle though angled

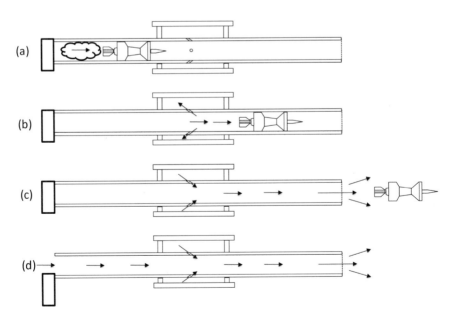

**Fig. 6.2** The mechanics of gas flow into, and out of, the fume extractor

ports and, (d) the breech is opened, and air is bulled from the turret to purge the bore. These are used in big guns and can be found on naval guns as well as tank guns. With tank guns, it is common to find that they sit asymmetrically around the gun and this is to avoid fowling on the tank's structure.

## 6.5  Liquid Propellant Guns

During and just after World War II there was a level of excitement for the prospect of using liquid propellants to propel projectiles. This was due to new developments in rocketry and the Space Race that was evolving at the time. The V2 developed in Peenemunde by Werner von Braun used a combination of ethyl alcohol and liquid oxygen as the propellant. The Saturn V moon rocket used a combination of kerosene and liquid oxygen in the first stage booster and used liquid hydrogen and liquid oxygen in the second and third stages. Such propellants are called cryogens because at least one of the components needs to be liquified at room temperature. However, rockets and guns have very different requirements when it comes to propellants. There were several attempts to develop liquid propellant guns after the war and patents started to appear in the 1950s and 1960s (Klein 1985).

One of the more recent developments that has occurred in the past 20 years was the development of a liquid propellant for artillery. The main attraction of using liquid propellants lies in the fact that logistically it would be easier and simpler to transport the energetic component to the frontline when compared to solid propellants. Further, it has the capacity to allow for an infinitely variable charge and therefore multiple firing solutions in terms of projectile velocity and hence range. However, there are several disadvantages in liquid propellant approaches. These include the fact that the liquid propellant would need to be fed to the breach under high pressure and therefore any risk of a leak could be catastrophic. Further, it necessitates large and complex breach mechanisms to accommodate the feed of the liquid propellant.

The US army trialled a system in the early 1990s in what was called the Advanced Field Artillery System or 'Defender'. Defender was the US Army's 155 mm liquid propellant gun developed by General Electric. There were over 125 test firings of this gun. Keeping the optimum volume void ratio at all angles of elevation was key to the success of this project and unfortunately lead to a complicated breech chamber system. As seen previously, the volume void ratio is a key parameter that defines how quickly the propellant burns. Too large a void ratio and the propellant will undertake a slow deflagration process, too small and there would be a rapid burn process. However, it was claimed that the basic operation of the Defender gun was quite simple and comprised of two main moving components that dynamically controlled the void ratio. These were the control piston and the injection piston (see Fig. 6.3). These two pistons moved within a breach plug. Both pistons moved forward during loading and rearward during firing. To begin the firing cycle, the required volume of propellant was pumped into the breach between the control piston and the injection piston. A specially designed ignitor was used to ignite the propellant starting the deflagration

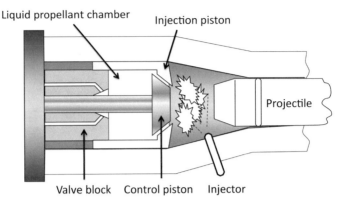

**Fig. 6.3** Schematic of a liquid-propellant breech system *Source* author

process. The pressure of this initial combustion caused the control piston to move rearward and create an opening between the two pistons. The propellant was then continuously injected into the combustion chamber through an annular orifice with the pressure increasing as more propellant was pumped into the chamber and burned. This process pushed both pistons rearward until all the propellant was consumed. There were perceived advantages of the system at the time and these included two important factors:

1.  Regenerate control (that is, the controlled movement of the pistons) of the entire combustion process provided for a soft launch for potentially complex artillery shells that used terminal guidance. Such shells usually employed complex electronic systems on board that would be sensitive to the forces from high accelerations.
2.  Precise measurement of propellant volumes made it possible to control the velocity and therefore improve accuracy.

Further, the propellant had a low vulnerability to ignition from a shaped charge. Data showed that when an 81-mm shaped charge jet was fired into identical containers of liquid propellant and water, it was apparent that there was no significant reaction with the liquid propellant despite the high shock pressure induced by the shaped charge jet. Nevertheless, Defender was ultimately cancelled.

## 6.6  Explosives

The most famous explosive is TNT – standing for trinitrotoluene. This was first discovered by the German chemist Julius Wilibrand (1839–1906) in 1863. Much of the work on high explosives was being powered by a renaissance in the understanding of chemistry and several advances were to be made in the subsequent 40 years or so. However, it turned out that Wilibrand's TNT was not originally used as an explosive material as it was too difficult to detonate. Oddly, it was originally used as a yellow dye. Its explosive properties were not full realised until it was studied by another German chemist, Carl Häussermann, in 1891. Häussermann was the first to appreciate the military potential of TNT and proposed its use as part of a gun-fired munition (Krehl 2008). The fact that it is hard to detonate meant that it could withstand the gun-shock during projectile launch. It was adopted by the German Army in 1902 and by 1914 was the most widespread energetic component of any exploding shell. Its low melting temperature (80 °C) lends itself to being melted and cast into shells. TNT is so important that ever since the first nuclear explosion in 1945, large explosions have been correlated in terms of TNT on what became known as a 'TNT equivalence'. This convention intends to compare the destructiveness of the event with that of TNT. Even release of energy through natural events such as earthquakes, volcanic eruptions, landslides and asteroid impacts, have all since been described in terms of a TNT equivalent mass.

Due to its inherent stability, TNT was extensively used in World War II and many of the shells contained TNT occasionally mixed with wax. Other explosives used included Amatol (a mixture of ammonium nitrate and TNT), PETN and Cyclonite (e.g., see (Bergin et al. 1953)). Cyclonite became known as RDX which was abbreviated from 'Research Department eXplosive', a name coined by the British. RDX was first prepared in 1899, and like TNT and guncotton, its initial use was not for explosive purposes (Akhavan 1998). It was discovered by Georg Friedrich Henning (1863–1945) of Germany and patented in 1898. Henning was a pharmacist and naturally his invention was intended for medicinal purposes. Ironically, probably unknown to Henning at the time, his invention was set to take lives, rather than preserve them. Developments in RDX proceeded in the US, Canada and the UK. However, it was not used as a main filling in British shells (unlike some German shells) but was added to TNT to increase the power of the explosion. RDX was mixed with TNT to form the charges used in the bouncing bombs dropped on German hydroelectric dams in Operation Chastise in May 1943. It was used in explosive compositions in Germany, France, Italy, Japan, Russia, USA, Spain and Sweden (Akhavan 1998).

Many large guns are designed to launch projectiles that contain sizeable amounts of high explosives such as TNT. These explosives are used in the propulsion of high-velocity fragments or to damage structures by a blast wave. Notably, the rate of energy release in a detonation is incredibly quick and occurs within the micro-second time-scale—much faster than gun propellants. Further, the detonation products are highly compressed gases that are rapidly deposited into the atmosphere to produce shock waves in air.

Pure explosives are very difficult to handle and so these are blended with other explosives or inert materials to change the mechanical behaviour or sensitivity of the material. Typical forms of products include: castings, slurries and gels, putties, machined polymer-bonded forms and rubberized materials. Explosives are separated into two types. There are the 'primary explosives' which are sensitive to flame/heat, percussion or friction. Then there are the secondary explosives which are less sensitive to external stimuli but reliably detonate to produce damaging shock waves.

It turns out that the amount of energy that is released in detonation is not especially high. This might seem an odd thing to say given the devastating effects of explosions. However, there are two features of high explosives that make them so dangerous. The first is that that the speed at which the energy is released during detonation is extremely rapid. It is the rapidity of the energy release that is important here not the magnitude. Secondly, the products that are formed during detonation are gases in an extremely compressed state. As these gases expand they are able to perform 'work' on their surroundings. It is the work on the local environment that we see when buildings collapse, craters are formed and so on.

How explosives detonate is still not completely understood and is based on the theory that a 'hot-spot' acts as the origin for the detonation. These hot spots are very small—generally of the order of microns in size. They cannot be accurately controlled, and neither can we predict where they will erupt in the explosive. All we can do is speculate that this is the mechanism by which explosive materials detonate. This is quite surprising really given the tonnes upon tonnes of explosives that have been used even over the past year by mining companies, as well as the military.

The reason why scientists cannot get to the bottom of the true nature of hot spots is because they are so tiny and they appear and disappear in an instant. So, they are really difficult to visualise in an opaque explosive, for example. However, there are really good theories that explains how explosive materials detonate. The first attempt to get to grips with this was done by David Chapman (1869–1958) in around the end of the 19th C (Chapman 1899). Importantly, there was a growing understanding that as an explosive detonated, it did so by the propagation of a wave, that Chapman termed an 'explosive wave'. We understand this today as being a 'detonation wave'. Chapman carried out a rather mathematical analysis to explain the results of some experimental data reported by a Professor Dixon in 1893. Similar theories were derived by the Frenchman, Jacques Charles Émile Jouguet (1871–1943) in 1905 (Jouguet 1905) and together they have been honoured by being named in a point that is used to explain the mechanism of detonation. The Chapman-Jouguet point (or CJ point) is a physical location in the explosive at which the solid explosive is transformed from solid into gas during detonation. It moves at the detonation velocity into the explosive (which is very fast).

The Second World War drove an enormous amount of innovation and no-more so than in weapons technology. Consequently, there were further advances on our understanding of detonation during the war that were proposed independently by three Russian, American and German scientists: Zel'dovich (1914–1987), John von Neumann (1903–1957), and Werner Döring (1911–2006). They independently proposed a model for detonation based on a Rankine-Hugoniot relationships developed in the 19th C (Cooper 1996). In their model, a thin shock wave compressed the explosive to a high pressure and temperature, which became known as the von Neumann spike. This pressure spike is finite in duration and very thin. The spike initiates an exothermic chemical reaction that is completed almost instantaneously. The energy liberated from the reaction drives the detonation wave. Simultaneously, the pressure and temperature is released from the heights of the spike and reduce to

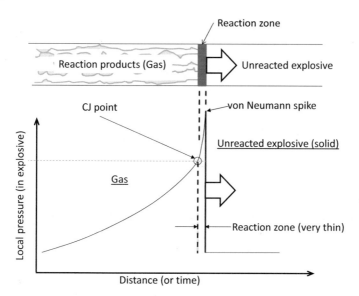

**Fig. 6.4** The detonation process showing the transit of a shock wave into the unreacted explosive (adapted from (Meyers 1994))

a pressure at the CJ-point, from which the gases formed during the chemical reaction expand, and the pressure and temperature are reduced. If we put a gauge in an explosive to see what happens to the explosive during detonation, the gauge would report a response as seen in Fig. 6.4.

When detonation happens, a chemical reaction occurs in the explosive that results in the expulsion of a gas at such a high rate that a shock wave is formed in air. The local air is being overdriven that forms the shock wave in a similar way in which a fast-moving supersonic jet leads to the formation of a shock wave. It is this shock wave that can tear flesh, propel fragments and lift vehicles. Quite a sizeable volume of gas is generated during the detonation. Roughly speaking, a 1 kg mass of TNT will produce 750 litres of gas as a result of the explosion.

The physics of blast waves is actually quite complicated however there is one golden rule for maximising protection against explosive devices: maximise your distance from the explosion. The reason for this is that the blast resultants (pressure and impulse values) decrease very rapidly as the distance from the explosive and the structure that you wish to protect increases. There is not a lot you can do to protect yourself when you are close to a detonating explosive. However even small distances of separation from you and the explosive charge can result in a better outcome in terms of survivability. This is the lesson that has been adopted by many armoured fighting vehicle manufacturers over the years where the blast is deflected from the base of the vehicle and the ground clearance is maximised. Other things can be done to protect from an explosive blast too. In the vehicle context, this can include blast chimneys where the explosive shock wave and 'blast wind' is vented through an internal channel in the vehicle. Recently, an interesting development has appeared that uses rockets on the roofs of the vehicle to provide a downward push after an explosive mine has detonated. This has the advantage such that the occupants of the vehicle are not subjected to huge 'bone-crushing' accelerations.

### 6.6.1  Unintended Detonation

Every now and again there is an explosive event that captures the eyes of the World. These include such devastating events such as the bombing of the Alfred P Murrah building by Timothy McVeigh and his co-conspirators in 1995 or the attack on the World Trade Center on 26 February 1993. As I compile this book, there has been an explosive event that devastated a large area of Beirut. On the 4th of August 2020, 2,750 tonnes of ammonium nitrate appeared to detonate and devastate the local port area. Such unintended large explosions in this day-and-age is unusual—mainly because much is known about how to safely store explosive products. And, due to the fact that many people had their mobile phones trained on the preceding fire that led to the detonation, lots of video footage exists.

For things to go very bad very quickly, there are a number of things that need to happen. And these, very sadly, happened in Beirut in August 2020. The blast registered a magnitude of 3.3 on the Richter implying a mini-Earthquake had occurred. A large red/brown column of smoke emanated from the site of the blast—confirming what was already known, that ammonium nitrate was the source of the explosion. Ammonium nitrate will release a lot of oxygen when detonating and when mixed with the nitrogen in the air, produces nitrogen dioxide, $NO_2$. Nitrogen dioxide has a distinctive brownish hue to it. Using social media, estimates of the blast equivalence in terms of the mass of TNT were between 500 kg and 1120 kg (Rigby et al. 2020).

Similar explosive events have happened before and sadly will probably happen again. In 1947 The S.S. Grandcamp—carrying around 2,100 tonnes of ammonium nitrate along with fuel, and ammunition—arrived at the Houston-area port with a fire in the cargo hold. By all accounts, the subsequent explosion killed some 600 people, including everyone on the dock and ship. Further there were more than 5000 injuries and the blast wave levelled 500 homes. This was one of the largest non-nuclear explosions in history—only to be surpassed, by some estimates, by the Beirut explosion.

In terms of gun use, literally millions of explosive shells have been fired from guns over the years. However, from time to time problems do happen. When you are firing an energetic substance from a gun it is important to realise that the type of explosive and the way that it has been filled into the shell, will affect its safety. So for example, we find that when inertial forces caused by the acceleration of the shell up the gun barrel causes one part of the explosive to move at a different rate within the shell to another portion of the explosive, then initiation can occur. This becomes a very dangerous scenario. The reason being is that when a shell explodes within a gun barrel the enormous pressure that results will cause the gun barrel to fail. And, this will result in metal fragments being thrown great distances. It turns out that there are several tests that can be performed that will provide data on the resilience of an explosive composition when it is subjected to such conditions and therefore most military grade explosive go through a rigorous safety testing process before fielding. Furthermore, there has been a push particularly in Europe and the US of late to develop an insensitive munition protocol such that it is actually very difficult for projectiles containing explosives to detonate when they are not expected to. For end users, this must be somewhat reassuring.

# References

Akhavan J (1998) The chemistry of explosives, 1st edn. The Royal Society of Chemistry, Cambridge, UK

Bergin WM, Collins JL, Thiebaud KE, Vendenberg HS (1953) German explosive ordnance (Projectiles and Projectile Fuzes). Departments of the Army and the Airforce, Washington

Chapman DL (1899) VI. On the rate of explosion in gases. The London Edinburgh and Dublin Philosophical Magazine J Sci 47(284):90–104. https://doi.org/10.1080/14786449908621243

Cooper PW (1996) Explosives engineering. Wiley-VCH, New York

Goad KJW, Halsey DHJ (1982) Ammunition (including grenades and mines), Vol III. Battlefield Weapons Systems and Technology. Brassey's publishers Ltd., Oxford

Hunnicutt RP (1988) In: Firepower: a History of the American Heavy Tank. Presidio

Jouguet E (1905) Sur la propagation des réactions chimiques dans les gaz [On the propagation of chemical reactions in gases]. Journal de Mathématiques Pures et Appliquées. 347–425

Klein N (1985) Liquid propellants for use in guns—a review. US Army Ballistic Research Laboratory, Aberdeen Proving Ground, Maryland

Krehl POK (2008) History of shock waves, explosions and impact: a chronological and biographical reference. Springer, Berlin Heidelberg

Meyer R, Kohler J, Homburg A (2007) Explosives, 6th edn. Wiley-VCH, Weinheim, Germany

Meyers MA (1994) Dynamic behaviour of materials.Wiley, Inc., New York. http:\\doi.org.10.1002/9780470172278

Munn OD, Wales SH, Beach AE (1857) Gun cotton and collodion. Sci Amer 13:109. https://doi.org/10.1038/scientificamerican12121857-109

Oxley JC (2003) Chapter 1—A survey of the thermal stability of energetic materials. In: Politzer P, Murray JS (eds) Theoretical and computational chemistry, vol 12. Elsevier, pp 5–48. https://doi.org/10.1016/S1380-7323(03)80003-1

Perrett B (1987) In: Soviet armour since 1945. Blandford Press

Rigby SE, Lodge TJ, Alotaibi S, Barr AD, Clarke SD, Langdon GS, Tyas A (2020) Preliminary yield estimation of the 2020 Beirut explosion using video footage from social media. Shock Waves. https://doi.org/10.1007/s00193-020-00970-z

# Chapter 7
# The Evolution of Projectile Design

Projectiles (or bullets) come in all shapes and sizes. With small hand-held guns, the bullets tend to be manufactured from several materials and are usually solid. Whereas, with larger and higher calibre guns, such as artillery guns the projectile will comprise of hardened steel casings containing a mass of explosive. Occasionally, gun-launched projectiles contain a shaped charge that can form and project a super-plastic jet of copper up to velocities of 12 km/s when the projectile strikes the target. However, these systems are less common in gun-launched weapons these days and tend to be reserved for shoulder-launched anti-tank weapons, air-to-ground missiles and torpedoes.

The word 'ammunition' is the term given to describe a bullet and its associated components that give it velocity. So, in the case of small arms and most cannon systems, the ammunition comprises of:

(a)  The propellant—the chemical fuel that can drive the bullet forward;
(b)  The primer—this ignites the propellant;
(c)  The cartridge case – this contains the propellant and primer;
(d)  Any ancillary pieces (such as propellant packing); and
(e)  The projectile (bullet).

Small-arms ammunition are typically characterised by the fact that the projectiles are small (relatively speaking) and the gun can be carried in an individual's 'arms'. A typical standard-issue 5.56 mm projectile is quite tiny, being only a few millimetres in diameters and around 22 mm long. The bullet generally consists of a penetrating mass (i.e., the bit that does all the work during penetration) surrounded by a gilding jacket that acts as a barrier, protecting the core of the bullet from the rifling of the barrel (and vice versa). Other types of projectiles include tracer and/ or incendiary rounds where an energetic material is integrated into the projectile design. The tracer located at the rear of the projectile is designed to be ignited by the propellant upon launch to provide a visible trace. The incendiary composition sited at the front of the projectile is designed to ignite upon target impact. The bullet itself comes in all sorts of shapes and sizes. Most bullets possess an ogival-shaped nose simply for

© Springer Nature Switzerland AG 2021                                             115
P. J. Hazell, *The Story of the Gun*, Springer Praxis Books,
https://doi.org/10.1007/978-3-030-73652-1_7

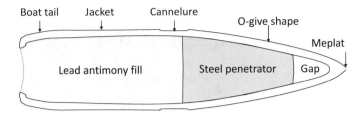

**Fig. 7.1** Schematic of a 5.56-mm M855/SS109 projectile showing a two-component core with lead antimony and steel components *Source* Author

aerodynamic stability and to reduce drag during flight. Similarly, some bullets may have a 'boat-tail' (that is, a slightly reduced diameter at the base) to reduce drag from the rear of the bullet (see Fig. 7.1). Often a small meplat (French = *méplat*, meaning "flat") is incorporated onto the front of the projectile to help with stability as well as improve terminal ballistic effects for certain soft targets.

We usually describe ammunition in terms of the bullet's calibre and cartridge length. So, ammunition that is described as 7.62 × 51 mm refers to a bullet of calibre of 7.62 mm and a cartridge length of 51 mm. The actual diameter of the bullet will be slightly larger than the stated calibre (by ~ 0.2 mm); and this is so that the bullet's jacket can engage in the grooves of the rifling, with the calibre being measured to the 'lands' of the rifling in the gun barrel.

As we have seen previously, the purpose of the rifling is to provide stability can be imparted to the projectile by causing it to spin during flight. As the projectile is fired, the rifle lands are driven into the jacket material. The rifling in the gun barrel is designed with a specific twist such that as the bullet moves along the gun barrel, it is forced to turn—thereby inducing its spin. However not all guns have rifling, as we have seen in the previous chapters.

## 7.1  The History of the Bullet

The modern English word 'bullet' is derived from the French word 'boulette' which means 'little ball' or 'pellet'. Historically, most of the bullets that were fired from guns were usually spherical in nature and mostly made from lead. As a shape, balls can maintain a reasonably straight trajectory and experience reasonably low drag forces compared to say, rectangular, oval or even cubic fragments and therefore it is not surprising that this was the desired shape for all projectiles fired from muskets. The word 'ammunition' also has a French origin. It was probably an altered form of the obsolete French word "la munition" which was used to describe all material that was used in war. Eventually it became a word that was used to describe powder and shot.

Most early projectiles consisted of lead shot and this was because lead is dense and has a low melting temperature and so can be easily formed into shape. Prior to gun-fired projectiles, Hellenistic slingers would be able to launch lead oval-shaped projectiles over large distances. They were thought to be more attractive than bow-launched arrows in that they were cheaper and inflicted more damage to the enemy. Lead melts at 327 °C whereas iron, which was another early contender for musket shot, melts at 1538 °C. In fact, melting lead could be done on a kitchen stove when cooking the evening's dinner. Its softness also lends itself to penetration of flesh as it easily deforms on impact and during penetration, causing disruption to internal organs. And, it was more forgiving to the internal surfaces of the musket barrel than iron. Iron is hard and if lodged in a gun barrel would lead to a potentially dangerous barrel failure.

To mass-produce lead shot it was common practice to use a shot tower. A shot tower was a tall chimney-like structure in which molten lead was sieved at the top and allowed to drop to the bottom into a water bath. The molten lead 'rain' was pulled into spherical shape by surface tension. In fact, if I asked you to draw a picture of a raindrop almost certainly you would draw a tear-shaped structure. The reality is that when water falls through the air, as raindrops do, they form a spherical shape. The reason for this is due to the surface tension. Lead has a much higher surface tension than water and therefore in a molten state, forms perfect spheres. This 'falling-drop' method was patented by William Watts of Bristol in 1782 and legend has it he stumbled upon his invention whilst witnessing a church fire. The fire melted lead from the roof and was solidified as it cooled and finally quenched by a puddle.

Lead of course is still used in projectile designs but the shapes have evolved. One of the first and most notable attempts to migrate to a projectile with an o-gival shape was with the invention of the Minié ball. This was a mostly conical bullet of soft lead invented by Captain Claude Minié of France (1804–1879) in the late 1840s. By adding more mass by virtue of length it was found to cause a more severe injury in the recipient, when compared to a spherical musket ball. It readily deformed and fragmented as it struck and tumbled through the body causing very nasty injuries. It was used extensively in the American Civil War (1861–1865) and the Franco-Prussian War (1870–1871) and was fired from rifled guns. The spin stabilisation and the stream-lined shape meant that the projectile would travel much further than the traditional musket balls.

Figure 7.2 shows a schematic of the Minié ball. The hollow at the rear was a means of providing obturation as the black powder burnt and caused the lead projectile to expand and press against the inside of the barrel, thus providing a seal. The cuts in the projectile towards the rear allowed for the projectile to find its path by allowing the rifling lands to easily embed into the relatively thin rings. Literally, these were so it would 'find its groove'.

Minie´ is properly pronounced 'min-yay'. Just so you know.

Most early bullets were very simple, and their complexity was limited to spherical lead balls as stated earlier. Lead, of course, is soft and small particles of lead would accumulate in the barrel during each firing. This would lead to barrel fouling thereby disrupting the path of a subsequent projectile. This was not ideal and required a different solution. That solution lay in the full metal jacket bullet that is commonly

**Fig. 7.2**  Claude Minié's
'Minié Ball' ammunition
*Source* Author

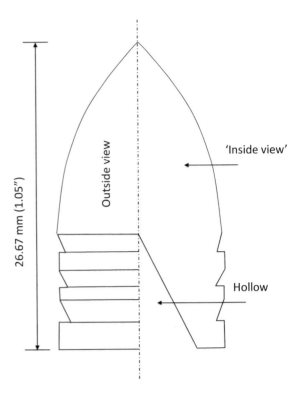

seen today and can date its history back over 100 years. This was invented by Colonel Eduard Rubin (1846–1920) of the Swiss army who was after a bullet that was both 'gentle' on the gun, that could fly straight, travel long distances and still have significant terminal ballistic effects at the target. In fact, bullets have actually changed very little since the work of Colonel Ruben when he invented the full metal jacket projectile in 1882.

## 7.2  The Role of the Bullet

Projectiles are accelerated up the gun barrel by the fast-burning propellant that applies a force on the rear part of the bullet. Remembering that Newton discovered that when a force is applied to a mass it will be accelerated (F = ma) and therefore the force that is applied by the pressure of the expanding propellant gases against the projectile's base area will cause the bullet to accelerate. In fact, high-velocity rifles can subject the bullet to enormous accelerating forces. As we have seen, we often refer to acceleration in multiples of the gravitational constant, that is g = 9.81 m/s$^2$. Bullets in high-velocity rifles can reach accelerations of 100,000 g. This is around

5000 times the acceleration that a human being can withstand. Therefore, small-calibre bullet designs cannot be too complex. The components of a bullet are usually limited to two or maybe three materials.

The high acceleration induced on launch can be problematic if a projectile is fitted with electronic components. These are more commonly found on higher-calibre projectiles such as the 155 mm Excalibur projectile which has the amazing ability to steer its way to the target. These projectiles are generally fired from systems such as the M109 Paladin self-propelled howitzer. The steering of a 155 mm projectile is no mean feat and can only be achieved by the clever deployment of fins. The gun-shock that is subjected to the Excalibur projectile is around 15,000 g—quite less than a high-velocity rifle bullet but nonetheless, potentially troublesome for the delicate GPS system, small electronic motors, and computer that manages the on-board systems. The novelty of the Excalibur projectile lies in its terminal guidance. The gun can be angled up to 20 degrees off target and the projectile still find its way to the target. In fact, the gun can be fired at up to 40 km from the enemy and still hit a target the size of an average bathroom. The terminal guidance that is provided by the fins enables an almost-vertical descent that limits collateral damage and improves localised lethality. So how do you protect the delicate components? Well, there are three principle ways of doing that. Firstly, you modify the propellant so that you use a relatively slow-burning propellant[1]. This gives the advantage that the peak pressure that is seen by the rear of the projectile is lower in order to provide more of a slow push to the base of the round rather than a fast 'jab'. This is most commonly achieved by using different structural shapes for the propellant grains and by incorporating a burning-rate modifier into the chemistry of the propellant. Secondly, you use glue! Yes, lots of glue! Basically, you coat the electronics components with a resin that effectively locks down the non-moving components. This is particularly useful with printed circuit boards and components that are soldered and stops any potential wiggle of the components. Thirdly, you employ lightweight and strong materials that can withstand huge loads such as titanium. You know that if in the Marvel superhero movie Ant-man (2015) Hank Pym, played by Michael Douglas, struggled to get through a hijacked Soviet missile shell made from 'solid titanium' then it has got to be strong! Titanium and its alloys are well known for low density and high strength. In fact, titanium alloys can withstand stresses of up to 1,000,000,000 $N/m^2$ before they start to flow plastically. This is basically equivalent to over 40,000 sports utility vehicles stacked up on top of each other on a square-metre plate.

An important part of bullet construction is the fact that there needs to be a good seal between the bullet's jacket and the inside part of the gun barrel. This is called 'obturation'. This was first suggested in 1832 by Captain John Norton. Norton's design featured a hollow base which expanded under pressure from the burning propellant to engage with the barrel's rifling. Apparently, the British Board of Ordnance rejected

---

[1]I know that slow-burning seems an unusual term—particularly as propellants are known to expire in several 1000th of a second. However, it is a term that is commonly used to describe a 'lazy' propellant. That is, one that takes more time than the average to reach the 'all-burnt' position in the pressure—time curve.

it because it contradicted their fairly fixed views on what a bullet should look like. That is, it was not spherical! Norton's design was also conical which broke from tradition.

Obturation is important in modern weapons and this is achieved by the way the bullet is made and the major diameter of the bullet. As previously mentioned, bullets for small arm guns tend to be oversized. So, a bullet with a 7.62 mm calibre would typically have a maximum diameter of 7.82 mm. Recall that the caliber of a rifled gun is measured from lands to land and therefore the oversized diameter bullet is squeezed in between the lands of the rifling as it is spun along the gun barrel.

For large weapons systems such as artillery guns it would be quite normal to employ a copper obturator that sits towards the rear of the shell. The purpose of that would simply be to act as a seal. For tank guns, polymer driving bands are often used. Uniquely for rifled tank guns where the projectile is to be drag-stabilized[2] (such as the APFSDS projectile used in the Challenger II tank) a slipping driving band would be used to obturate the propellant gases. The reason that it should slip would be limit the spin that would be imparted to the projectile, although notably a small amount of spin is desirable simply to cancel out any inbalances that occur due to the manufacturing process.

## 7.3  The Role of the Jacket

The jacket not only provides a good seal for the projectile as it travels along the gun barrel. The jacket serves an additional three purposes: (1) to protect the barrel from the core; (2) to engage with the rifling of the barrel and (3) to provide a projectile shape optimised for free-flight. For example, with tungsten–carbide-cored bullets such as the 7.62 mm M993, the jacket metal is made from soft steel that envelopes a tungsten carbide core with an acute front tip.

The bullet's jacket can affect penetration too—particularly into hard-faced armour such as those manufactured from ceramic. It has been known since 1878 that cast iron projectiles would not penetrate a dual-layer armour plate where the impact surface was hard but when fired at the softer rear surface of the same plate (i.e., after rotating the armour plate), perforation occurred. This resulted in a Captain English proposing that a projectile with a 'cap' of relatively soft wrought iron would stop the projectile from shattering to secure perforation against a hard face. Sadly, English' proposal was not exploited at that time (Johnson 1988). Later it was proposed by Stepan Makarov (1849–1904), a Russian Admiral, in 1893. Makarov realised that if a ductile cap was added to the shell, then the shock of the impact would be diminished thereby protecting the penetrator. It also had a secondary effect in that it provided confinement to the tip of the shell thereby hindering fractured material in the projectile separating and therefore enhancing the penetration. By 1915, this approach had been integrated

---

[2]That is to say, fins are employed to keep the projectile pointing in the trajectory of flight and therefore stabilize it.

into most Armour-Piercing shell designs (Anon 1915). It was an approach that was also used extensively into World War II by both sides of the conflict and was the technique applied with the British 6-pounder Armour-Piercing Cap (APC) projectile and Armour-Piercing Cap Ballistic-Cap projectile (APCBC), for example.

With modern jacketed projectiles the impact situation becomes slightly more complex in that hard-cored armour-piercing projectiles are designed to attack hard ceramic-based armour systems where both the core of the projectile and the ceramic plates are intrinsically brittle. So, in some ways the jacket is acting like a shock absorber when the bullet strikes these really hard armour targets (similar to the Makarov cap). However, it also can stop the core of the bullet from radially expanding when it collides with the target and keeping everything as narrow as possible leads to better penetration. We call this process 'inertial confinement' in that the jacket is stopping the projectile's core from rapidly expanding.

## 7.4 Soft Bullets

The role of the bullet is often not as clear as one might think. Of course, they are designed to penetrate. But penetrate what? For soft targets (i.e., human beings) the projectile needs to be reasonably soft and that is why 'ball' projectiles are manufactured with lead cores (a soft metal) so as to deform at low impact pressures. Very often the element antimony is added to alloy with the lead so that the core is a bit stronger than the strength pure lead would provide. So, the core would be manufactured from an alloy of lead and antimony. The amount of antimony is usually limited to around 5%. The core would then be cased in a copper or brass (a mixture of copper and zinc) jacket. Lead, of course is not very good for you and in high doses can attack the nervous system and the brain which can in very rare circumstances lead to coma and death. Impact within a confined firing range can produce fine lead-based aerosols that can be harmful to health. Children are the most vulnerable. These factors have created the need for non-lead projectiles.

The purpose of a ball-type bullet is to transfer as much energy into the human body as quickly as possible. And, this is achieved by the rapid expansion of the bullet core as it begins to penetrate and then decelerate. As the bullet expands, the bullet comes into more contact with stationary tissue and it is this that causes it to slow down. Thus, kinetic energy (the energy that is derived from movement) is transferred into the tissue of the poor subject that has been struck. When softer tissue is struck by a modern high-velocity Full-Metal-Jacket (FMJ) bullet it may not disintegrate but rather tumble—at least initially. The tumbling effect of the bullet can be particularly nasty for the incumbent as the muscle, organs and other soft tissue is crushed and torn to bits. Notably, some bullets will tumble more than others with shorter bullets being more stable in human tissue than longer bullets. We will look at this again in the next chapter.

The question that follows from this line of reasoning is how do scientists *know* how bullets behave in human tissue? Well, the answer is pretty straightforward and involves the study of a number of sources. Firstly, of course there are the cadavers. However not many people these days offer their deceased bodies for target practice in the name of military science. However back in the 19th C. that did not appear to be a problem and in fact Theodore Kocher carried out his research using human cadavers between the years of 1875–1900 and much of this work has formed the basis of understanding of how bullets interact with human flesh. Nevertheless, a more modern and palatable test material is ballistic gelatin. Ballistic gelatin is essentially jelly (i.e., the type of material that one can easily access at the supermarket); but it is deliberately calibrated so that depth of penetrations measured experimentally correspond to that seen in the human body. Other possibilities include ballistic soap. This is quite like your ordinary bathroom soap with the superb benefit in that unlike gelatin, it will retain permanent deformation after being penetrated. The advantage of the gelatin is that it provides a translucent view of the bullet penetrating and thereby provides an idea of the temporary cavities that potentially occur.

More recently the types of structures that one sees having potential for tissue simulants are materials such as porous polyurethane. One material that has been used in this regard is commercially known as Synbone. This material has curiously been designed so that surgeons can practice cutting and manipulating bone and has similar mechanical properties to that of human bone. However, it is not clear how similar this material is under ballistic impact loading conditions.

## 7.5   Hard Bullets

For armour-piercing bullets, the purpose is quite different from the ball bullets previously described and these (as the name implies) are designed to penetrate through hard plates. These projectiles are made with very hard and strong cores. So, again, the usual design of a jacket envelope that contains a hard and pointed core. The cores of these projectiles need to be strong and so are made from either high-strength steels or materials such as tungsten carbide.

Although there had been various 'hard' projectiles that had been used against wooden Naval ships, the first of the modern armour piercing projectiles was invented by Sir William Palliser (1830–1882) in 1864, and so named after him as the Palliser shot. This was a cast iron elongated projectile that could be used in muzzle-loading rifled cannons. The nose of the projectile was chilled so as to harden it so that it would resist deformation and this allowed for better rigid-body penetration of hard targets. A cavity was provided in the middle to mitigate against cracking that would otherwise occur when casting a solid shell (Cundill 1877). Bronze studs around the periphery facilitated engagement with the rifling. Later, 'studless' Palliser shells were introduced for 9-inch and 12.5-inch calibers and fitted with an obturating gun-metal (brass) plug at the base. Eventually, driving bands were fitted to the rear, similar to modern shells. In time, they were also capped to allow for better penetration of hard targets.

During the Second World War, the Germans adopted an "arrow-head" design that was developed in 1939 by Polte in Magdeburg. These were used in every German anti-tank gun from 1942 onwards (Backofen Jr and Williams 1979). The Soviets copied the design in the development of their "spindle" projectile (see Fig. 7.3). There was one difference. The German arrowhead projectiles provided obturation at the front shoulder whereas for the Soviet spindle projectile, this was provided for at the rear. Front obturation led to less balloting during launch and allowed for a longer projectile leading to better penetration. The German arrowhead projectiles used magnesium or aluminium windscreens possibly to achieve better behind-armour-effects but more likely to provide sight of a flash when the projectile struck the target (Bergin et al. 1953). The Soviets also developed aluminium windshields for their spindle projectiles, possibly for the same effect. Both magnesium and aluminium alloys exhibit pyrophoric behaviour (pyrophoric = Gk for 'fire-bearing'). That is, they tend to spontaneously ignite when smashed to very small pieces in an oxygen environment. Sometimes, this can add to the destructive effect after the core has made its way through the armour.

The armour-piercing anti-tank kinetic-energy projectiles used during World War II usually made use of a tungsten carbide core. Tungsten carbide is around twice the density of steel and is a hard strong ceramic-like substance. It is a compound

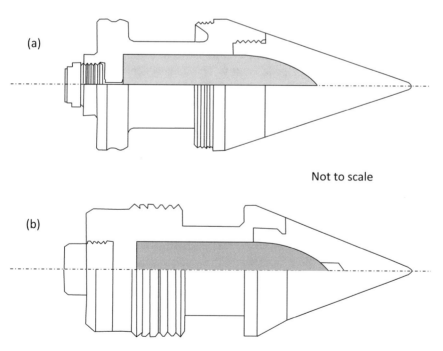

**Fig. 7.3** Armour Piercing projectiles: **a** 47-mm German arrowhead projectile (Panzergranate Patronen 40) **b** Soviet "spindle design", largely based on the German arrowhead projectile. The grey section represents the tungsten carbide core

comprising of tungsten and carbon atoms and is often manufactured with another material such as iron or cobalt that acts as a binder to provide toughness. Its density means that for the same volume of bullet the weight of the bullet would be double that of a steel core. This means that the kinetic energy of the projectile (which depends on the mass of the bullet) is higher.

Tungsten carbide is commonly used on blades used for cutting materials that are used in workshops and the like and is frequently used in mining. It was also used during World War II by the German Luftwaffe in the ammunition for their notorious tank-killer: the Stuka JU-87G. It was also used in anti-tank munitions by the American and British forces and is still used today in some bullet designs such as the 7.62 × 51 mm FFV, the 7.62 × 51 mm APS9 and the 14.5 × 114 mm BS41.

So, what determines whether a hard projectile will penetrate through an armour plate? Well, there are several factors. First there is the kinetic energy of the projectile. Armour works by either absorbing the kinetic energy of the projectile during the process of crater expansion or by destroying the projectile and thereby causing the fragments (and kinetic energy) to disperse outside of the armour. Consequently, the *kinetic energy* of the projectile is incredibly important. Remember that the kinetic energy of a projectile is determined by the following equation:

$$KE = \frac{1}{2}mv^2$$

where m = the mass of the projectile and $v$ is the velocity of the projectile. The velocity term has a power term (i.e., = 2) and therefore doubling the velocity leads to a four-fold increase in kinetic energy. This means achieving a high projectile velocity is very important for penetration. The Germans knew this back in World War I where they used the charge from a 13-mm caliber machine gun cartridge to fire a 7.92 mm bullet.

This is true for almost all materials—but not all. If the velocity is too high, the bullet core could be shattered by the sharp shock pressure that is subjected to the bullet core when it strikes the target (hence the need for the Makarov cap on early designs). By way of example to illustrate this, if you were able to fire a hard and strong tungsten carbide sphere at a thin Perspex[3] target at, say 2 km/s, the Perspex target, although being defeated by the projectile, would induce such a large *shock pressure* into the projectile that it would cause it to shatter. The shock pressure occurs by virtue of the fact that large shock waves are induced into both the target and projectile simultaneously and it is these shock waves that cause extensive fragmentation damage to the projectile. Of course, the target would also be completely obliterated during this process. Now of course, all projectiles that are launched from propellant-based guns are travelling at less than 1500 m/s with most travelling at ~ 850 m/s. Nevertheless, no armour is made from Perspex.

---

[3]Perspex is a plastic that has a hard and brittle appearance but like all polymers the mechanical properties are an order of magnitude less than that of tungsten carbide.

Mass is also important and that perhaps explains why some projectiles are made with heavy cores (particularly tungsten carbide). However, mass is not as important as velocity. The other factor that affects how a projectile penetrates a target is the cross-sectional area. So, a projectile with a low cross-sectional area will provide enhanced ability to penetrate compared to one with a larger cross-sectional area. However, the requirement for high velocity and low cross-sectional are contradictory. Reducing the diameter of the projectile reduces the area over which the propellant gases can 'push' the projectile along the gun barrel. There were a couple of ways around this conundrum that were developed during World War II. The first was by having a projectile with a large surface area at the base that was reduced as the projectile accelerated towards the muzzle. Thereby it was possible to achieve a high velocity with a relatively small cross-sectional area as it left the muzzle. As we have seen in Chap. 3, this is the type of approach that was used by the Littlejohn squeeze-bore shot. This had two flanges that were swaged down by an attachment fitted to the end of gun barrel. However, the amount of 'squeeze' that you could be applied to the outer diameter of the projectile was limited and the barrel wear could be quite excessive due to the large contact forces between the barrel and the parts that were being squeezed down. Commonly a rifled section towards the breech end conveyed the spin to the projectile whilst at the muzzle-end the bore was smooth. These tapered bore guns were used in World War II although they had been suggested as early as 1903 by Karl Puff and again in 1929 by Gerlich. The German 42 mm PAK 41 anti-tank gun used this approach. The projectiles included the use of high-explosive-filled shells and armour piercing projectiles with a tungsten carbide sub-projectile with smaller calibres of 28 mm (squeezed to 20 mm) and 42 mm (squeezed to 28 mm—see Fig. 7.4) also being used (Bergin et al. 1953). These projectiles could be launched to 1100–1400 metres-per-second, which is comparable to the performance of today's guns (Backofen Jr and Williams 1979). The approach became known as the 'Gerlich principle'. The concept is shown in Fig. 7.5.

The second approach that is mostly commonly applied today makes use of a sabot. The word sabot comes from the French word for shoe or boot. Originally, sabots were a type of footwear that were made by hollowing out a block of wood

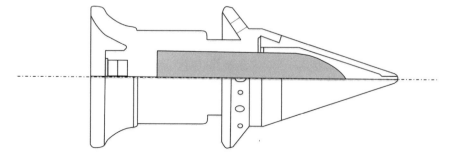

**Fig. 7.4** Armour-piercing projectile: Tapered bore gun projectile (42–28 mm); Panzergranate Patronen L. Panzerabwehrkanone 41 (Pak 41)

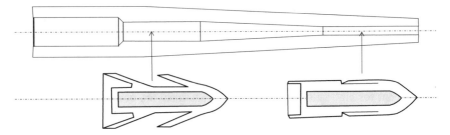

**Fig. 7.5** The squeeze-bore shot concept (adapted from (Germershausen 1982))

to securely receive the foot. They were traditionally worn by French peasants. Of course, in munition design they are intended to securely support the projectile as it travels along the gun barrel. Their advantage is that they offer a larger cross-sectional area to the expanding propellant gases to carry the smaller diameter 'sub-projectile' along the barrel. After the projectile leaves the gun barrel the separation of the sabot from the penetrator allows for a streamlined projectile to travel to its target.

There were several sabots that were used as projectile supports in the 19th C. In Chap. 4. we saw that a sabot system was also suggested by a Dr von Eberhard in 1916 as a method to increase artillery ranges beyond 100 km with the intention of shelling Paris, and later London. For the Paris-gun project, von Eberhard's suggested a 35.5 cm gun could launch a sub-calibre 21.0 cm projectile, that was spin stabilized and carried by a discarding sabot. Historically, this was the first suggestion of a spin-stabilized sub-calibre, saboted projectile. However, von Eberhard's solution was not adopted by the makers of the massive Paris gun at the time, which was arguably a bad decision. Instead, a high working pressure was adopted by accelerating a 106 kg, 21.0 cm full-calibre shell to 1600 m/s through a 21 cm rifled tube (and smooth bore extensions). This decision curtailed the shot-life of the barrels and therefore limited the degree of bombardment. Arguably, this decision, in part, has allowed us to still enjoy Paris' ancient beauty to this day (Bull and Murphy 1988).

Importantly for the British, a modern sabot design was introduced just before World War II by engineers working for the Edgar Brandt company—a French company. And, much of the work continued during the war with Permutter and Coppock as we saw in Chap. 3. (Buckley 2004). The main problem for the British was that they had limited weapons with which they could penetrate the German tank armour. In addition, the design of the British tanks in the 1940s precluded the instal-lation of a higher-calibre and longer gun. Early projectile designs comprised of a relatively small diameter tungsten core which was carried by lightweight aluminium collars, the outer diameter of which matched the bore diameter of the gun. This was called Armour-Piercing Composite Rigid (APCR) shot. This approach kept the collars in place as the projectile flew through the air. However, the projectile lost velocity and hence the lethality effect at the target was diminished. Consequently, Permutter and Coppock modified the design by engineering weak points into the aluminium sabot parts and thereby allowing them to fail when subjected to aero-dynamic forces. The goal was so that the sabot elements would literally peel away from the tungsten core as it travelled towards its target without disturbing the core's trajectory. This round was named 'Armour-Piercing Discarding Sabot' shot (APDS).

The German Army also began work on a sabot projectile around a year before the end of the war. Most of their work was carried out at the Hillersleben Proving Ground and discovered by the Allied forces at the end of the war. The attempt there was to find the most efficient and economical way of discarding the sabots (Bergin et al. 1953). It seems, they were unsuccessful, however.

The trick with using a sabot is to ensure that the amount of force that is provided by the expanding propellant gasses used to accelerate the sub-projectile is as large as possible. So, the intention is to not waste propellant energy propelling the sabot. That is why modern sabots used in tank guns (where the velocity of the penetrator is very important) are made from low density materials. More recently they have even employed carbon fibre composites to make the sabot parts (e.g., see Fig. 7.6). Although, magnesium and aluminium alloys have also been used due to their low density.

There are also several ways in which the sabot is connected to the sub-projectile and this has been largely determined by the type of sabot and the materials of the sub projectile. For long penetrators made from tungsten alloys or depleted uranium—which are weak but dense materials, it is common to use a large array of connecting threads that mate the sabot to the sub projectile. These threads are constructed in a

**Fig. 7.6** The carbon-fibre sabot used as part of 120-mm APFSDS tank ammunition
*Source* Author

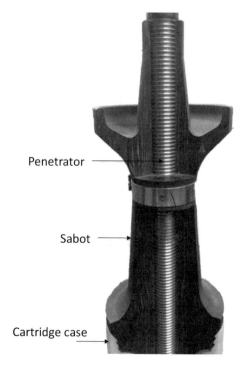

Penetrator

Sabot

Cartridge case

geometry of a buttress shape and are either machined into the sub projectile or they are sintered in place.

## 7.6   Long Projectiles

Having small cross-sectional area has many benefits for a projectile, not least the fact that the drag forces on it during flight are reduced compared to larger diameter projectiles; and therefore, they can maintain high velocities to a farther distance. Some of the early projectiles that made use of this of this phenomenon were termed 'arrow' projectiles. It seems ironic that the medieval weapon of choice for ranged attack was also an arrow; and we have almost come full circle in our projectile design. In fact, some of the earliest reports of gun-fired arrows date back to 1356 in Korea where an arrow was fired 'some distance' and buried itself in the ground up to its feathers (Chase 2003).

There is some evidence that the Germans were developing Fin Stabilized Discarding Sabot (FSDS) projectiles before and during World War II (Backofen Jr and Williams 1979). These projectiles resembled an arrow in shape. By all accounts, their sabots were proven in the 88-mm smooth bore barrels. However, the Germans demonstrated better success with larger calibre high-strength bunker busters. The first was the Röchling projectile (see Fig. 7.7a). The unusual characteristic of this projectile was the fact that it used fins that were flexible steel sheets that were wrapped around the rear of the projectile. These fins unfurled after the projectile left the muzzle and allowed the projectile to be drag-stabilized—much in the way that the common dart is before it strikes the dartboard. The Röchling projectile was an early type of bunker buster designed specifically for large howitzers where the projectile was lobbed into the air and onto its target. The largest of the shells was for the gigantic 35.5 cm Haubitze M1 siege howitzer. The projectile alone weighed in at 1000 kg and was expected to penetrate over 100 m in soil.

The Peenemünde projectile (Fig. 7.7b) was another attempt at an arrow-type projectile which carried an explosive payload and was fronted by a solid steel penetrator element. Like the Röchling, it was terribly inaccurate but boasted an awesome range of 90 km when fired from the 170-mm K5 Long Range Gun. With this gun, the muzzle velocity reached 1850 m/s and could reach altitudes of 130 km when fired vertically. This meant that it could pass the Karman line implying that it could be fired into outer space almost reaching the altitude for low earth orbit! In fact, as we have seen in Chap. 4, this was the dream for the famous ballistics expert Gerald Bull and Project HARP (High Altitude Research Project). A 105 mm version of this projectile was found after the war among German experimental work. The design of this projectile, with its discarding sabot ring, inspired British developments in the late 1940s (Holton 1949). These projectiles were fin-stabilised and it is not clear if the intention was to design an armour-piercing variant or develop an explosive-carrying projectile in the same vein as the Peenemünde projectile. Nevertheless, there are clear similarities between the German, US and British developments in the 1940s

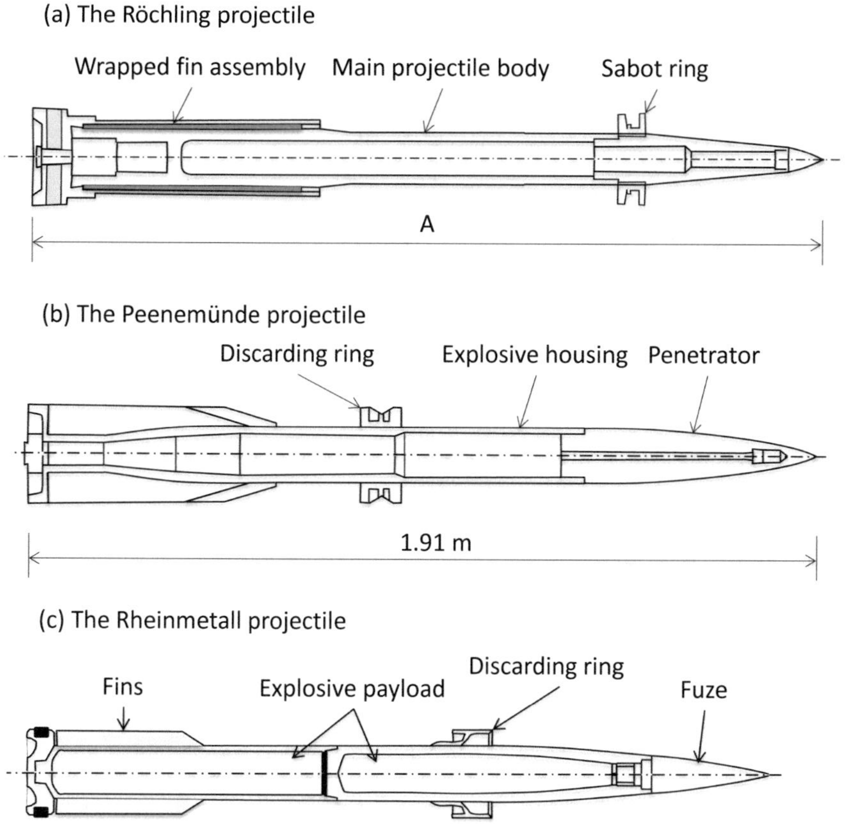

(a) The Röchling projectile

Wrapped fin assembly    Main projectile body        Sabot ring

A

(b) The Peenemünde projectile

Discarding ring        Explosive housing   Penetrator

1.91 m

(c) The Rheinmetall projectile

Discarding ring

Fins          Explosive payload                 Fuze

**Fig. 7.7** Three German arrow projectiles: **a** The Röchling projectile showing the contained fins; A = 2.2 m for the 210-mm caliber Howitzer gun and A = 3.6 m for the massive 355 mm siege howitzer and, **b** The Peenemünde projectile showing the main penetrator part and the zone for the explosive payload, **c** The Rheinmetall long range projectile. All images are sections

and 1950s to the designs deployed by the Soviets in the 1960s and 1970s, such as the BM series of APFSDS projectiles of which the BM-15 is an example. See Fig. 7.8b.

The Peenemünde projectile was also the inspiration for the Rheinmetall long range projectile that was developed in the 1960s (see Fig. 7.7c). This projectile was developed for 155 mm artillery system and could be fired from both rifled and smooth-bore tubes. These projectiles allowed for an increase in range by 30% compared to 'standard' 155 mm shells. However, there was a penalty in the explosive payload. Nevertheless, the accuracy was regarded as being excellent (Germershausen 1982).

Many modern-day tank guns fire sub projectiles that look like darts. These are called Armour-Piercing Fin-Stabilised projectiles or 'long rod penetrators' and I am going to refer to them as LRPs from now on. The fins themselves generally account for around 20% of the total projectile drag at supersonic speeds and are consequently highly swept back in design to limit this drag (e.g., see the Peenemünde projectile

(a) 1949 – British Experimental Shell FSDS 5.4 / 2.48 – 966 mm long (5.4 in / 137 mm calibre)

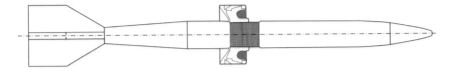

(b) 1972 – Soviet Union – BM 15 – 547 mm long (125 mm calibre)

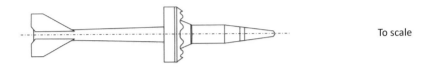

To scale

(c) 2003 – US M829A3 – 930 mm long (120 mm calibre)

**Fig. 7.8** From FSDS to APFSDS. The shell at the top of the image was an experimental projectile developed and trialled by the British in 1940s using a 5.4-inch gun **a** (Holton 1949). Section of the sabot is shown. This was inspired by a 105 mm Peenemünde projectile. This in turn probably lead to the inspiration of the BM series of projectiles such as the BM-15 projectile introduced by the Soviets in the 1970s based on previous models **b**. Both have bore-riding fins and short sabots. The bottom image is the current state of the art showing a section of a projectile that employs a DU penetrator with a carbon-fibre sabot **c**

above in Fig. 7.7). This concept was pioneered in 1935 by the German scientist Adolf Busemann (1901–1986) and led to improvements in supersonic aircraft design in the 1940s and 1950s.

LRPs have a characteristically higher velocity than high-velocity rifle bullets and can reach velocities of up to 1800 m/s. Further, the calibre of the gun is much larger too being typically between 105 and 125 mm (see Fig. 7.8 b). Some cold-war-era experimental tank guns had calibres of up to 152 mm, with the MBT-70. Again, velocity is very important. However, the velocity of the projectile has a limited effect on the penetration. That is to say, there is a velocity above which it is not possible to get more penetration for a given length of projectile. This is perhaps counterintuitive as we are putting more energy into the target–right? Correct. However, as an example, for a tungsten-based LRP, the maximum amount of penetration into armour steel that could be expected would occur if the impact velocity was between 3.0–3.5 km/s. And, in fact (and perhaps insanely strange) theoretically it would not matter too much if we were able to increase the velocity to 10 km/s (if that was possible). The penetration depth in the target would be about the same. This weird phenomenon is called the 'hydrodynamic limit' and is all to do with the way the projectile penetrates at these elevated velocities. The truth is that at these higher pressures the penetrator and the target behave in accordance with the laws of fluid dynamics (hence 'hydrodynamic'). Therefore, the penetration rate is constant, and the rod is continuously eroded as it penetrates. The projectile is defeated when it is mostly eroded, revealing a short rigid body penetrator with a large blunt nose. This section is brought to rest. The reality is that during penetration, the projectile behaves as a super-plastic solid. So, the metal is flowing under huge stress created by the force of impact. Materials that ordinarily would fail with strains of 30% can withstand strains of 500% in this scenario. This is all because of the way the large compression experienced by the projectile during penetration limits void growth and crack formation and this allows for the material to flow freely without 'fear' of failure. Our understanding of this phenomenon was first developed in the 1960s independently by the British scientist, Tate (Tate 1967) and the Soviet scientist, Alekseevskii (1966).

As the projectile is only defeated when the length has been completely eroded, it stands to reason that the length of an LRP is also important. The longer the penetrator the more penetration that could be expected before it is consumed. Modern LRPs such as the M829A3 approach 1 m in length (see Fig. 7.8 c). It also stands to reason that the higher the density of the penetrator the greater the inertial forces acting on the target material. So, there are two key properties of LRPs that are required for maximizing penetration: length and density. The length maximizes the time in which the projectile can penetrate the target by virtue of the fact that it is being continually eroded in the process. Whereas the density of the penetrator provides more 'push' into the target material. Early examples of penetrators, such as the Soviet BM-15 (Fig. 7.8 b), were made from high-strength maraging steel body and employed a tungsten carbide 'slug' at the tip. This allowed for a shorter discarding sabot as the steel material had sufficient shear strength to accommodate the high g-forces during launch with a limited number of threads between the sabot and sub-projectile. However, due to the density of the steel (7800 kg/m$^3$) penetration depth was limited to less than its length. That is why tungsten and depleted uranium were later introduced into modern designs. Both materials have very high densities; but of course, they are not nice materials to handle. The other disadvantage was that the materials' relatively low shear strength dictated much longer sabot sections to mate with the rod.

Depleted uranium or 'DU'[4] is uranium that has had some of its U-235 content removed and therefore is only weakly radioactive. In fact, the alpha particles that are emitted by the material will only travel a few centimetres in air and can be stopped by a sheet of paper. So, holding a DU projectile will not affect your health. However, if the projectile is broken down into small parts and somehow ingested into the body then the alpha particles can damage the sensitive tissue inside the body. This will lead to an increased risk of cancer and other diseases. However, that is not the only problem as DU is extremely toxic—especially to the kidneys and the lungs. This was a fact that appeared to be lost on the scientists of the mid 19th C. who thought

---

[4]Most modern DU projectile materials are alloyed with a small (0.75%) amount of titanium to improve strength and corrosion resistance.

that uranium possessed homeopathic properties, being particularly useful for treating diabetes!

The idea of using uranium in projectiles is thought to originate with the Nazis. Albert Speer (1905–1981) served as the Minister of Armaments and War Production in Nazi Germany during most of World War II and he was a close ally of Adolf Hitler. Therefore, he would have had a handle on the use of uranium. He notes:[5]

*In the summer of 1943, wolframite imports from Portugal were cut off, which created a critical situation for the production of solid-core ammunition. I thereupon ordered the use of uranium cores for this type of ammunition. My release of our uranium stocks of about twelve hundred metric tons showed that we no longer had any thought of producing atom bombs.* (Speer 1970)

This account was largely treated with some skepticism. However, evidence of Nazi involvement in the nuclear arms race has since been uncovered; and therefore, it stood to reason that they had access to large uranium supplies that could otherwise be used for shells (e.g., see (Bernstein and Cassidy 1995)).

In fact, I have been provided with a photo of a round that was recovered from a basement of a University in England (Backofen 2020). The uranium-based ammunition was 110 mm long from the tip of the penetrator to the base of (what looked like) the cartridge. The cartridge had a 29 mm diameter at the base. The diameter of the penetrator was approximately 16 mm. Apparently, it was radioactive (that is, not depleted).

During the 1991 Gulf War it was thought that around 300 tons of DU was deposited into the ground in Iraq and Kuwait. Although tank guns fire DU armour-piercing projectiles the DU was mostly delivered by the famous A-10 tank busters. These aircraft carry 30 mm Gatling-type guns that can fire DU anti-materiel projectiles at 1000 m/s and at up to 4200 rounds per minute. So, it is no surprise that many of those cores were shattered when the projectiles struck the metal or concrete targets. The shattered core material would have inevitably been broken down to extremely fine particles and therefore posed a risk of inhalation as an aerosol. DU particles are heavy (uranium is the eighth densest naturally occurring element) and so the particles would quickly settle to the ground where there is a possibility that it would find its way to the water table, and, in the end drinking water. So why is DU used in projectiles? Well, the short answer is that it is cheap! For example, the US government has ~ 500,000 metric tons of the stuff in surplus storage as a byproduct of nuclear processing. Sure, you could use gold, rhenium, platinum, iridium or osmium, all of which are denser than uranium. However, you would need VERY deep pockets to equip an army! Secondly, it is also has an improved performance compared to its tungsten counterpart in that its alloys have a slightly higher density. This results in a higher penetration. It also has a remarkable ability to self-sharpen during penetration by virtue of the fact that DU is susceptible to shear failure. This sets up failure planes set at an angle from the contact point between the LRP and the target. This results in the penetrator maintaining a sharp point throughout the penetration process. The inevitable consequence of this is that it is easier to penetrate the material due to

---

[5]In Albert Speer's 1969 memoirs, translated into English in 1970.

larger contact pressure being maintained in the target. Thirdly, it is pyrophoric and will burn rapidly at relatively low temperatures (150–175 °C) and spontaneously ignite at temperatures of 600 °C. Therefore, the bits from the LRP that make it through the target are set alight by the high contact stresses and the frictional forces that are present spewing hot and fiery fragments everywhere. Nasty!

Many countries do not use DU and choose tungsten as the base metal of the projectile that is used in tank guns. However, tungsten is hardly innocent either. Tungsten is usually mixed with other elements to alloy the material to achieve optimum material properties. Examples include tungsten-nickel-iron alloys and the tungsten-nickel—cobalt alloys and it is the presence of the nickel and cobalt that provides the potential for the cancer. All in, weapon design is a nasty business!

## 7.7  Explosive Projectiles

In 1784, Lieutenant Henry Shrapnel (1761–1842) invented a hollow round shot that contained a fuze, a powder bursting charge and several small pellets that were spread over the battlefield. This was a revolutionary invention because for the first time, multiple troops could be engaged by a single projectile. Of course, the pellets became known as 'shrapnel'—a term used today to describe fragmented metal from an exploding projectile. There have since been lots of innovative and deadly examples.

One of the more unusual gun-fired projectile designs fielded by the German infantry in World War II were the super-calibre 'stick grenade' projectiles, or 'stielgranate'. There were two types. A shaped charge projectile (the Stielgranate 41) that was deployed with the 3.7 cm Pak 36 guns and a high-explosive type (the Stielgranate 42) that was launched from the 15 cm Sig 33 (schweres infanterie geschütz 33 = heavy infantry gun 33). The Sig 33 gun was built by Rheinmetall (a company that is in production today and has a wide portfolio of Defence equipment that is sold to many countries). The Sig 33 was developed in the early 1930s and early models were made incorporating aluminium-based alloys. Most of the Sig 33's projectiles were quite conventional—except for the Stielgranate 42. The Stielgranate 42 was a super-calibre explosive projectile, that naturally was muzzle-loaded. It was fitted with a driving shaft and was principally an explosive shell for achieving demolition, clearing minefields and the like (see Fig. 7.9). The overall length (including the shaft was) 1.28 m.

Many of the artillery projectiles that are fired these days contain significant amount of high explosive that produce large fragmentation effects and a devastating shock wave that can pick up objects and propel them sizeable distances. Explosive shells were commonly used in tank guns as early as the 1950s in the form of a High Explosive Squash Head projectile (or HESH). In fact, at the time, it was thought that this would be the projectile to 'end all' projectiles—especially against concrete bunkers. It was not designed to penetrate the armour. Instead, it worked by delivering a layer of explosive to the outside of the target that spread (or 'squashed') into a shape that resembled a cowpat. And so the projectile was a carrier projectile that simply

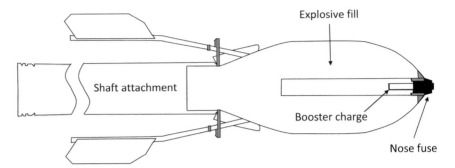

**Fig. 7.9**  The Stielgranate 42 projectile showing the shaft attachment  *Source* Author

allowed for the spread of the explosive on the outside of its intended target. Next, the explosive was detonated when the fuse, carried in the rear part of the projectile, came into contact with the cowpat of explosive and detonated it. The result was a powerful shock wave that travelled into the target. The shock wave travelled faster than the speed of sound in the material and so it did not take long to reach the rear surface of the target. Now, the sonic velocity in a material depends largely on the proximity of the atoms within the material, stiffness of the bonds between them and the mass of the atoms. So, for example, the speed of sound in steel would be close to 6000 m/s whereas in concrete it is around 3400 m/s. Therefore, the shock wave would traverse an armour-plate in a tiny fraction of a second. When it encountered the rear of the target the compressed moving volume of material driving the shock wave would be 'reflected' back into the compressed material as a damaging tensile wave as the rear free surface began its motion. As a result of the tensile wave the material begins to split along a plane of the target's surface until a large fragment separates from the inside. This fragment can travel at around 130 m/s—which is approximately one-third the speed of sound. In fact, in steel quite large fragments could be 'ejected' and so in the 1960s, it was thought of as a good option to defeat the Soviet tanks. HESH was first introduced as tank ammunition with the 120 mm L1 gun of the heavy Conquerer tank and also with the 105 mm L7 gun of the Centurion. However, it is easily defeated by spaced and layered compound armour.

## 7.8  The Science of Exterior Ballistics

Bullets, shells and penetrators need to fly through the air and this explains their shape. In essence, they tend to be stream-lined as much as possible. Bullets and shells need to be constructed in a way that their bases are thick and strong to withstand the heat and pressure in the breach. This has a tendency of pushing the centre of mass rearward and therefore the centre of pressure, which is the centre through which all the drag forces act, will sit forward of the centre of mass. Thus, the only way these projectiles can be stabilised is through spin stabilisation. On the other hand, if projectiles are long and thin, where the centre of all of the lift and drag forces act behind the centre of mass, then the projectile can be drag stabilised. A common way of 'forcing' the centre of pressure behind the centre of mass is by using fins at the rear to increase aerodynamic forces acting on the relatively large surfaces that are provided by the fins.

For nearly all handguns, artillery guns and some tank guns, stabilisation[6] is achieved through rifling. The purpose of the rifling is to impart powerful gyroscopic forces into the projectile. In its simplest form, a gyroscope can be described as a heavy inverted cone, like a child's spinning top. When balanced on its apex and stationary it simply falls over. However, when it is spun, it is able to stand up. So, spinning a projectile allows for it to remain stable as it moves through the air. However, there is a limit to this and it is determined by the length-to-diameter ratio of the projectile. As soon as that ratio approaches ~7, we find that the projectile would have to be spun insanely fast to maintain stability. To test this, try spinning a pencil so that it remains upright!

Inducing spin into a projectile for the purpose of improving accuracy had been known about since the 15th C. Rifling of gun barrels is thought to have originated in Augsburg in Germany around the year of 1498. However it did not become mainstream until the 19th C. Arguably the earliest and most complete treatise on the benefits of rifling we given by Robins in 1742 (Robins 1742) which was republished in 1805 due to the work being 'much desired by the public'. Benjamin Robins (1707–1751) was regarded as Britain's *Father of Gunnery* and Germany's *Founder of Experimental Aerodynamics*. He was a contemporary of Isaac Newton who died in March 1727 and caught the attention of Henry Pemberton who was the editor of Newton's 3rd edition of *Principia* in 1726 (Johnson 1992).

Around the mid-19th C., work was carried out to establish the best form of rifling in a gun barrel. However, many early examples of the guns that were made burst simply because the rifling introduced stress concentrations during the firing process. Further, it was found that if the force required to engrave the projectile was too great then the projectile would not progress along the gun barrel as quickly as anticipated. This increased the risk of the barrel bursting due to the volume not expanding as quickly as the pressure was increasing. Getting this right was a tricky problem to solve. Robins noted that this could be overcome by using leather that could be greased and wrapped around the lower part of the projectile and forced into the grooves of the gun barrel. That way, it was the leather that was engraving into the grooves rather than any metal of the projectile. This technique was employed in muzzle-loaders in Germany and Switzerland around the 18th C. Robins also noted that when firing ball projectiles, which was commonplace in the 18th C., that the projectile had a constant deflection to the left (early evidence of the 'Magnus effect'). Robins' remedy to the

---

[6]It is worth mentioning that some gun-launched HEAT rounds are 'shape-stabilised' and use a spigot that uses a small flange to invoke a rotating air pocket at the front of the projectile body. The length of the spigot and position of the flange are key to maintaining stability. Usefully, the spigot doubles as stand-off, the length of which is optimized for stability, not shaped charge penetration.

deflection was using bullets of an egg-like form instead of spherical. This resulted in a straighter flight-path.

### 7.8.1   The Magnus Effect

Even though Robins was also an early proponent of what became known as the Magnus effect, he was never credited for the observation. The conclusive description was published 100 years later (Magnus 1852) by Heinrich Gustav Magnus (1802–1870). It was Robins that conducted experiments with bent smooth-bore barrels and observed the counter-intuitive relationship that a loose spherical projectile would not follow the trajectory of the bend but rather deviate in the opposite direction (Johnson 1986). In other words, a barrel that was bent to the left would impart spin on the projectile as it contacted the right side of the barrel and that would result in it deviating to the right. He correctly noted that as the projectile was passing through the bent part of the barrel it was being forced to roll up on the side of the barrel and therefore a smooth bore barrel would impart spin. The spin affected the interaction with the air to the extent that as the projectile span some of the air is carried around with it due to skin friction. This results in one side experiencing less pressure than the other side. The same effect can be seen with golf balls when an undercut induces backward spin onto the golf ball. This results in lift and therefore a greater range. The same effect was seen by Newton with spinning tennis balls and is well-known to every soccer-player who tries to bend the football into the back of the goal net!

A spinning projectile can result in other effects that can affect a projectile's flight. One of the nuances of spinning gyroscopes is that as they are pulled in one direction they will react so that they will move in a direction perpendicular to the force with which they were pulled. This has a bearing on the flight trajectory of a spinning projectile as any dip in the nose or side movement due to a minor instability will lead to a perpendicular movement in the projectile. Thus, a spinning projectile is susceptible to what we call a 'nutation process'. This will cause the shells to precess around a trajectory. In this case the centre of mass of the shell will follow a trajectory whilst the nose will precess around that trajectory with an average offset to the right (for a clockwise stabilised shell). Therefore, even in completely still air, a bullet experiences small sideways movement due to its yawing motion. The yawing motion of the bullet means that the nose of the bullet points to a slightly different direction than the trajectory of the bullet. Thus, the bullet experiences a small 'side-wind' component of force. It is this side-wind component that results in lift or drop depending upon the spin direction.

An example of the Magnus effect is shown below in Fig. 7.10. Generally, the Magnus force is quite small compared to other forces that are acting on the projectile.

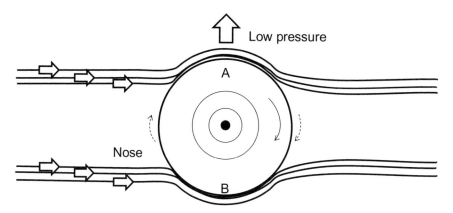

**Fig. 7.10**  The Magnus effect for a projectile with anti-clockwise spin  *Source* Author

## 7.8.2  *Projectile Drag*

All projectiles are susceptible to drag. As the projectile traverses the air, the air is displaced, and energy is required to displace that air. This energy loss is a continual drain on the kinetic energy of the projectile. This leads to a loss of velocity. The average of the drag forces acting at the front of the projectile is called 'forebody drag' and this increases dramatically as the speed of sound is approached. This is because the air can no longer escape to the sides from the front of the projectile and therefore subjects the projectile to an increasing force. The air bunches up to form a pressure wave or 'shock wave'. It is this pressure wave that can cause havoc to control surfaces for aircraft that are not designed to travel at supersonic velocities. During the Second World War, it was largely thought that humans could never exceed the speed of sound as pilots observed their instruments froze and the plane's structure vibrated heavily as the sound velocity was approached. Pilots spoke of a 'mysterious force' and it was as if they approached 'a brick wall in the sky'. Hence the speed of sound was called the sound 'barrier'. These myths were dispelled by Chuck Yaeger (1923–2020) as he flew his specially designed Bell X-1, dropped from a B-29 bomber in 1947. It was the design of the aircraft that allowed Yaeger to exceed the sound barrier and these lessons are also applied to projectiles too. Supersonic projectiles therefore tend to be sleek to minimise the effects of forebody drag.

Behind the projectile there exists a region of turbulence called the 'wake'. The air cannot fill the space behind the projectile quick enough and therefore a region of low-pressure forms. This results in turbulence leading to what we call 'base drag' which pulls on the projectile. Frequently boat-tailing was implemented to reduce this effect. This is where the rear of the projectile is shaped inwards to reduce the base diameter. For artillery shells an alternative approach to reduce the effects of base drag was developed by bleeding a gas into the low-pressure region behind the shell. This technique was developed in Sweden in the 1960s and a Swedish patent for it was filed in 1971. By the 1990s, the base-bleed technology was used extensively in a number of artillery shells. This provided artillery shells with a much greater range. Generally, gases of low molecular weight (e.g., helium) are more effective at increasing the base pressure than higher molecular weight gases (e.g., nitrogen or argon) (Tanner 1975).

There are other drag forces acting on the projectile too. These include skin-frictional forces due to the presence of friction between the skin of the fast-moving projectile travelling in the air. Another type of drag force is called excrescence drag. This is a result of protrusions that may exist on the projectile such as driving bands.

Air places an enormous hindrance on projectiles and as we have seen, this effect is exacerbated as the speed increases. So, for example take a 300 mm mortar shell with a muzzle velocity of 400 m/s. If it could be fired in a vacuum it would theoretically travel 16 km. We know that it only travels 11 km. Or take a 155 mm artillery shell which would have a muzzle velocity of 700 m/s, theoretically in vacuum would travel 49 km whereas we know in air it travels around 24 km at that speed. Increasing the muzzle velocity further, a 7.62 mm NATO ball round, with a muzzle velocity of 840 m/s in a vacuum would theoretically travel 70 km whereas in air the absolute maximum distance that projectile would travel would be around 3–4 km. So when comparing vacuum to air (and negating projectile size effects that also will play a role on the drag), the faster the velocity, the larger the reduction in range and this is explained by the increased drag.

### 7.8.3  And Underwater?

Projectiles travelling underwater are therefore also subjected to huge drag forces. A common question that is often asked is how far will a bullet travel underwater? Additionally, will firing a weapon underwater cause the barrel to break?

Firing conventional weapons underwater is never advisable. However, it is unlikely to cause the barrel to fail. Of course, water does not possess any stiffness or strength. But, we must bear in mind that every cubic metre of water has a mass of 1000 kg. Or put it another way, if you fill the volume occupied by a standard two-seater sofa with water—that would have a mass of approximately 1000 kg! That is why, for example, breakers in the ocean can knock you over when you play in the surf! So, for the gun to function underwater the projectile will need to shift the mass of water that sits in the barrel. That mass will not be much. However, the water in the barrel is further contained by the surrounding water. This is why it is difficult for projectiles to travel large distances underwater; and in fact, most pistol-fired projectiles are only likely to reach a couple of metres before coming to rest and sinking. Coming back to the ocean analogy, it is like trying to walk through water that is waste height. It is just more difficult to do than when we walk on dry land! In the barrel, any resistance to projectile movement as the propellant gasses expand and push on the rear of the projectile can lead to a pressure spike in the cartridge case. This is because the rate of volume expansion (due to projectile movement) is not sufficient to accommodate the expanding propellant gasses. To compound matters, the burning rate of the propellant is increased as the local pressure in the cartridge increases and this leads to an even higher pressure spike. This can, in rare occasions, lead to a weapon failure. And, this has sometimes resulted in injury. Most modern weapons have pressure-relief ports built into the chamber just in case there is a cartridge-case failure. However, these too will be impeded by the mass of the water. So, although it is unlikely to cause barrel failure, firing a weapon underwater can be dangerous and therefore not advisable. That is, unless you are a baddie in a crime thriller and your life depends on it! Even then, do not expect the projectile to travel far!

# References

Alekseevskii VP (1966) Penetration of a rod into a target at high velocity. Combust Explos Shock Waves 2(2):63–66. https://doi.org/10.1007/BF00749237

Anon (1915) Treatise on ammunition, 10th edn. War Office, London, UK

Backofen Jr, Joseph E (2020) Personal communication. Uranium penetrator on a 35 mm slide showing a scale bar. Canberra, Australia

Backofen JE Jr, Williams LW (1979) Soviet kinetic energy penetrators—technology/ deployment. Battelle, Columbus Laboratories, Columbus, Ohio

Bergin WM, Collins JL, Thiebaud KE, Vendenberg HS (1953) German explosive ordnance (Projectiles and Projectile Fuzes). Departments of the Army and the Airforce, Washington

Bernstein J, Cassidy D (1995) Bomb apologetics: Farm Hall August 1945. Phys Today 48(8):32–36. https://doi.org/10.1063/1.881469

Buckley J (2004) In: British armour in the Normandy campaign. Taylor & Francis

Bull GV, Murphy CH (1988) Paris Kanonen–the Paris Guns (Wilhelmgeschütze) and Project HARP: The Application of Major Calibre Guns to Atmospheric and Space Research. E.S. Mittler, Herford Germany

Chase K (2003) Firearms: A Global History to 1700. Cambridge University Press, Cambridge, UK

Cundill JP (1877) Treatise on Ammunition 1877. Secretary of State for War, London, UK

Germershausen R, (Ed.) (1982) Rheinmetall Handbook on Weaponry. 2nd edn. Rheinmetall GmbH, Germany

Holton RG (1949) Fin Stabilised Discarding Sabot Shell: Progress of Design Armaments Design Establishment. Ministry of Supply, Fort Halstead, Kent

Johnson W (1986) The Magnus Effect—Early investigations and a question of priority. Int J Mech Sci 28(12):859–872. https://doi.org/10.1016/0020-7403(86)90032-9

Johnson W (1988) Some conspicious aspects of the Century of rapid changes in battleship armours, ca 1845–1945. Int J Impact Eng 7(2):261–284. https://doi.org/10.1016/0734-743X(88)90029-2

Johnson W (1992) Benjamin Robins (18th century founder of scientific ballistics): some European dimensions and past and future perceptions. Int J Impact Eng 12(2):293–323. https://doi.org/10.1016/0734-743X(92)90486-D

Magnus G (1852) Über die Abweichung der Geschosse. Abhandlungen der Königlichen Akademie der Wissenschaften zu Berlin:1–23

Robins B (1742) New Principles of Gunnery: Containing the Determination of the Force of Gunpowder, and an Investigation of the Difference in the Resisting Power of the Air to Swift and Slow Motions. J. Nourse, London, England

Speer A (1970) Inside the Third Reich. Simon & Schuster, New York

Tanner M (1975) Reduction of base drag. Prog Aerosp Sci 16(4):369–384. https://doi.org/10.1016/0376-0421(75)90003-2

Tate A (1967) A theory for the deceleration of long rods after impact. J Mech Phys Solids 15(6):387–399

# Chapter 8
# The Science of Terminal Ballistics

Terminal ballistics is my favourite subject. I started my career teaching this as a young lecturer at the Royal Military College of Science at Shrivenham. In a nutshell, it is all about the science of how a projectile penetrates a target. This is clearly relevant to guns. The whole point of guns is that they deliver some form of terminal effect. And, when I mean 'terminal' I am referring to the fact that the projectile comes to rest in something or in someone. So, commonly the shooter will want to incapacitate or kill an individual or pierce or destroy a structure. Either way, it is a process of energy transfer. That is, the transfer of the kinetic energy of the bullet to the penetration of a material. So, now we will look at the science of terminal ballistics which encompasses the science of wound ballistics.

## 8.1 The Basics

As mentioned earlier, it is the kinetic energy of a projectile that does the work on the target (energy is sometimes described in terms of 'work'). Take the case of what happens when somebody is shot. As the bullet enters the body, some of that kinetic energy is used to pierce and break the skin, more still is used to separate the sub-dermal tissue and fracture bone. Some of the kinetic energy is consumed through the distortion of the projectile. This distortion process results in the projectile heating up. A small amount of heat will transfer to the surrounding tissue. More still is converted to the kinetic energy of the internal organs and perhaps the rupture of arteries. The kinetic energy is also being used to cause the separation of material from the bullet (= fracture).

© Springer Nature Switzerland AG 2021
P. J. Hazell, *The Story of the Gun*, Springer Praxis Books,
https://doi.org/10.1007/978-3-030-73652-1_8

## 8.2   Is Firing Guns into the Air Safe?

In many parts of the world it is customary for people to fire their weapons into the air during a time of celebration. This can be for reasons such as public holidays, presidential-election wins or even weddings. Ammunition can be cheaper than fireworks and in some people's minds, the thrill of firing a gun is just as exciting as launching a few fireworks. This raises the question as to whether a falling projectile from such a shooting is dangerous. There is certainly anecdotal as well as scientific evidence to suggest that.

Children are susceptible to serious injury and even death from such incidents (Hanieh 1971). The skulls of children are thinner than adults and thus less protective from falling objects. Take the tragic case of Marquel Peters. Just after midnight on the morning of January 1, 2010, four-year-old Marquel Peters was sitting next to his mother in a church in Decatur, Georgia, when a bullet perforated the building's roof and struck the small boy on the head. He died shortly afterwards. The shooter, it seems, was never found. And there are other anecdotal cases (Blanchfield 2015).

For the case of falling bullets, individuals will fall fowl of a gunshot injury even though they are not in the proximity of the shooter. For the case of Marquel Peters, ballistics experts speculated that the projectile, which was most likely fired from an AK 47, could have been fired anywhere from a half-mile to three miles away. In 1994, a research study reported 118 patients who had been treated for injuries after being hit by projectiles. Many of the victims had no idea of where the bullets had come from or who shot them. Some had not even heard a weapon fire. The sources of the bullets were identified in only six cases and in those six cases, the shooter was up to 1.5 km away. In those 6 cases, handguns were used. It was assumed that these casualties were the victims of gun fire from a vertically orientated weapon. The study reported that most [77%] were hit in the head with 32% of the victims dying as a result of their injuries (Ordog et al. 1994).

Many people believe bullets literally disappear when they shoot into the sky and therefore there will be no consequence. Some think that somehow the projectile makes it into space and stays there! As we have seen, many high-velocity projectiles have velocities in the range of between two-times and three-times the speed of sound. For something like a 7.62 mm-calibre projectile fired into the air vertically, it will reach a height of approximately 2.4 km in around 17 s. It will then take around 40 s to fall to Earth. Oddly, it will have the tendency to fall base-first as the projectile is more stable in rearward flight (even though the projectile is accordingly subjected to significantly more drag). The terminal velocity of such projectiles would be around 70 m/s. Given that the velocity of a projectile required for skin perforation is around 45 m/s and 60 m/s (based on penetration experiments into cadavers and a freshly killed pig (Haag 1995)) it is predictable that these falling bullets would cause serious injury and even death. Further, at the apex of flight (i.e., at the point that the projectile begins to fall to Earth) the wind can affect the return trajectory and so it is most likely for the bullets to fall to Earth between 1 and 2 km away. So, although it is very unlikely to cause injury to the shooter, a falling bullet may seriously injure or even kill an innocent bystander. So, do not do it!

This leads me on to the science of wounding.

## 8.3   The Science of Wounding

Much of our understanding of wound ballistics was developed through the work of Emile Theodor Kocher (1841–1917). Kocher was born in Bern, Switzerland in 1841 and became Professor of Surgery at the University of Berne in 1872. He was a deeply spiritual man as well as a brilliant surgeon. He won the Nobel prize in medicine in 1909, being the first surgeon to do so. That prize was for his pioneering work on the Thyroid. However, he was less known for his other considerable contributions and that was in the field of wound ballistics.

Around that time, the wounds from ballistic impact were becoming quite severe due to new projectile designs and in particular the proliferation of the Minié ball lead projectile. As discussed in Chap. 7, this was heavier than the previously-used musket ball lead projectiles and the resulting wounds were far more nasty. Furthermore, developments in nitrocellulose propellant, that was more powerful than the traditional black powder, also had a deleterious effect on wound morphology due to higher projectile velocities. Kocher's research had the objective to understand the mechanisms of wounding so that soldiers could have a meaningful recovery from their wounds.

At that time, it was widely thought that the wounding mechanism of tissue by a projectile was caused by three factors (Fackler and Dougherty 1991):

(1)   a partial melting of the bullet on impact;
(2)   the centrifugal force of the spinning bullet fired from a rifled barrel, and;
(3)   hydraulic pressure (that is to say, a high-pressure region emanating from the bullet and expanding out radially behind it).

Kocher, fired at several targets including metals, glass bottles, pig bladders and pig intestines and human cadavers. Most pertinently, he fired a lead-alloy projectile from a Vetterli rifle into a water-filled box faced with a pig bladder. The box was 1.5 m long. The pig's bladder covered a hole in the box. Kocher observed that the bullet pierced the pig's bladder, penetrated the water and struck the rear wall of the box. Almost instantaneously, the box burst catastrophically at the seams. Kocher concluded that hydraulic pressure (i.e., a high-pressure region radiating out from the bullet to form a temporary cavity) caused the box to burst and therefore was primarily responsible for wounding. Mapping his observations on to the penetration of the human body he concluded that the cavity that was formed was primarily responsible for tissue disruption. All other contributions were regarded as minimal. Consequently, Kocher concluded that partial bullet melting, and centrifugal forces were of little importance to wounding (Fackler and Dougherty 1991).

At the time of Kocher, 'fishing' with Dynamite was a common past-time and in 1898 it was shown in that firing a bullet close to a fish resulted in its death without any obvious injury (Times 1898). It was reported that an Italian Officer, Major Michelini fired an Italian (0.256-inch calibre) rifle[1] at water at an angle of 45 degrees resulting in a dead fish floating to the surface. It was deduced that the death of the fish was due to the hydraulic shock caused by the bullet penetrating the water. Although this phenomenon was referred to as 'hydraulic shock' the death of the fish would have almost certainly been down to the temporary displacement of the water behind the penetrating bullet. This was the same phenomenon that caused Kocher's box to burst.

There were other notable works in the injury mechanism at the time including the work by La Garde who had the macabre intent on investigating how a projectile's trajectory altered during penetration of human cadavers (La Garde 1916). La Garde would suspend cadavers in his care with tackle and shoot various projectiles at them using an array of contemporary pistols and rifles. There was heavy criticism of La Garde's technique as dead tissue was well-known to respond in a different fashion to living tissue when stressed mechanically. It was La Garde's intention to show that by comparing photographs and dissections of cadavers with data from the recent wars of the time that his study was a sensible approach. Nowadays, it is accepted that cadaver testing does provide valuable information regarding the response of the bone structure to impact and can be informative in understanding how loads change as they pass through bone. However, changes in tissue properties happen rapidly upon death and so it is difficult to get reliable ballistic data a projectile's effect on soft tissue (Young et al. 2015). Most modern work in this space, the reader will be pleased to know, is more directed towards the development of computational models to predict behaviour.

Possibly in a most bizarre set of experiments reported by La Garde was that 'poisoned projectiles', using anthrax could be used to infect the enemy. La Garde fired projectiles that were doped with anthrax from various types of handguns at 'susceptible animals' at a range of 500 yards. The majority of all his targets died from the anthrax disease. If anything, this showed that the bacterium survived the high temperatures of launch and the shock of the collision (see p 135, (La Garde 1916)).

Since the work of Kocher and La Garde, it is generally recognized that there are two principle modes of wounding due to projectile penetration. Firstly, there is a permanent cavity that is formed. This is simply caused by dynamic shearing or tearing of the tissue as the bullet pierces flesh. Clearly, if there are critical organs or arteries in the path of the projectile then very serious injury or death will result. The second type of wounding is due to the fact that a temporary cavity is formed—similar to the hydraulic shock that was mentioned previously. This is where local tissue is accelerated outward radially by the force of the penetrating projectile. This force acts locally for a short period of time. If however, an organ is not very 'elastic' (and

---

[1]Probably a 6.5 × 52 mm Mannlicher-Carcano rifle. This is the same type of rifle used in the assassination of President John F Kennedy.

there are several organs that fit this category, including the liver) then serious injury will ensue (Fackler et al. 1984).

It should be noted that there have been several myths that have evolved over the years that have been perpetuated by well-meaning scientists. These have been nicely summarised by Fackler (1988). So, let us do some 'mythbusting'.

### 8.3.1 Low-Velocity Projectiles Do not Cause Serious Injury

It is generally thought that the higher the velocity of the projectile the more injury is expected to occur. There is good reason for this. A higher velocity equates to a substantially higher kinetic energy (through the velocity-squared relationship). However, according to Fackler, there is therefore a widespread dogma that has propagated that high-velocity projectiles cause wounds that require extensive treatment, whereas those caused by low-velocity projectiles need little or no treatment. This is a false assumption, and the truth is far more complicated. It is known that slow projectiles can also cause serious injury. It is the degree of projectile fragmentation, deformation and interaction with inelastic organs that dictate the type of injury and whether debridement (that is, removal of damaged tissue) is necessary – not the velocity of the projectile. This can occur with low-velocity projectiles as well as high-velocity projectiles.

For example, a 1960s article suggested debridement was unnecessary for wounds caused by bullets with a muzzle energy that was less than 400 foot-pounds (542 J) (Morgan et al. 1961). That is around the muzzle energy of a low-velocity pistol bullet such as the 8.04 g (124 gr) Federal FMJ 9 mm Parabellum. It is difficult to believe that this will always be the case. Low-velocity rounds can also produce cavitation[2] and can cause shearing and tearing of tissue. Furthermore, this assumption also ignores the observation that serious wounding can occur if a buckle or button or other hard object is pushed into the victim. This can also result in substantial injury. (Hollerman et al. 1990).

### 8.3.2 Kinetic Energy Transfer Is to Be Viewed as a Wounding Mechanism

There have been several reports that claim that it is the kinetic energy transferred to the target that is the principal cause of wounding. Again, the truth is far more complicated and therefore this is not always the case. Fackler's main argument here was to do with the fact that quite extensive amounts of tissue in the human body respond elastically to a penetrating projectile. Therefore, tissue can behave like a

---

[2]Cavitation in this context is the formation of a temporary cavity in the human body due to the passing of a projectile.

rubbery substance when stressed. Of course, the temporary cavity that is formed can lead to blood vessel damage and even broken bones. However, wounds that result from projectiles with the same amount of kinetic energy can differ widely, depending on how the projectile interacts with the tissue and of course whether tumbling occurs.

### 8.3.3   Overemphasis of the Effects of the Sonic Pressure Wave

During penetration, a 'sonic pressure wave' will emanate from the tip of the projectile. This wave will 'ripple' out from the projectile tip at a velocity of around 1450 m/s in a similar fashion as you would expect if you dropped a stone into a pond. Whether this pressure wave causes injury has been the cause of an ongoing controversy. In 1947 researchers examined the effect of the sonic pressure wave on tissue by suspending frogs' hearts in a water vat and shot a projectile that passed near to them (Harvey et al. 1947). They recorded the experiment using high-speed photography. Since the sonic pressure wave travels ahead of the projectile, they were able to observe with the camera what exactly damaged the heart and separate out the effect of the hydraulic pressure from that of the sonic pressure wave. They noted that the disruption of the tissue accompanied changes in pressure only due to the formation of the temporary cavity and not the sonic pressure wave. This classic work has often been used to conclude that sonic pressure waves have no wounding effect.

Even though that the pressure of this sonic wave can approach 100 atm (~ 10.1 MPa) the wave is only thought to last 2 μs. A lithotripter that is used to treat kidney stones will generate a shock wave three times the amplitude of the sonic pressure wave from a penetrating small-arms projectile and up to 2000 of these waves are used in a single treatment session. However, with this more intense exposure of tissue to high pressure waves there is no damage to soft tissue surrounding the kidney stone (Fackler 1988).

### 8.3.4   Spheres Assumed to Be a Valid Model for All Projectiles

Many wound ballistic researchers have used spheres to evaluate the effects on the body. The reason for this is simple: many universities and research establishments prefer to operate small smooth-bore gas-guns which are unable to fire commercial ogive-shaped projectiles. Spheres offer some simplicity to the set-up in that they do not yaw or tumble. Commonly, they would be launched toward the target carried by a separating polymer sabot, some remnants of which invariably find their way to the target. Using a sphere, also makes computational modelling exercises much simpler too, which is generally important for academic publication. Further, many spheres are sold as ball-race materials that equally do not deform (unlike lead-based ballistic projectiles). All of these factors mean that wounding from a gun-fired bullet is going to be quite different to a gas-gun fired saboted sphere. Fackler showed that a penetrating sphere produces its maximum disruption near the target's entry, unlike gun-fired bullets. A pointed, non-deforming bullet causes the most disruption when its yaw increases the surface area acting on the tissue. Thus, the effect is starkly different.

### 8.3.5  Animals of 10–20 kg Assumed to Be Valid Models for Humans

Again, this is a misconception as it has been shown that where there is larger amount of tissue to move, the temporary cavity that is formed is smaller. This is due to conservation of momentum and will not only result in a smaller temporary cavity being formed but its expansion will be slower. Ultimately this could mean that misleading conclusions could be reached. As Fackler points out, that a bullet's size cannot be changed without affecting its ballistic and terminal ballistic properties and therefore there is no choice but to increase the size of the test animal to approximate the dimensions of adult humans if validity is to be maintained.

### 8.3.6  Use of Tissue Simulants with Unproved Equivalence to Living Tissue

As we have previously seen in Chap. 7, there are several tissue simulants that are available on the market and these range from ballistic gelatin (basically very similar in touch and feel to the jelly that you buy from a supermarket), ballistic soap and even a dielectric gel (e.g., Sylgard™) that reportedly behaves in a similar fashion to human brain. One is kind of left wondering how this likeness was stumbled upon! And then you have the bone simulants such as BoneSim™ and Synbone® and although there have been several attempts to compare these simulants to living tissue e.g., (Brown et al. 2019), there has been limited success. The truth is that living tissue is necessarily complex in structure. It has been designed that way (or evolved that way, depending on your philosophical perspective). Furthermore, dead tissue is no substitute either due to dehydration. I have had colleagues recall to me how there was a mad dash from the operating theatre to the mechanical testing lab, just so that the tissue that was being tested was fresh and as close to being hydrated and alive as possible.

## 8.4  Stopping Power

The amount that a bullet expands is indicative of its stopping power and its ability to incapacitate the enemy. This is 'polite' way of saying that the enemy combatant has been severely injured, perhaps even fatally, eventually. Equally, suppression of the enemy is often desirable. These two terms are defined below as:

1.  Incapacitate: 'A soldier is incapacitated when they are unable to carry out their primary task, at the required level of performance, because of being wounded'
2.  Suppress: 'A soldier is suppressed when they are unable or unwilling to carry out their task effectively, because of the actual or perceived threat.'

There is a case to be made that in a military context, it is desirable to wound your enemy rather than killing them. An enemy soldier that is disabled in action, such they are no longer able to fight, uses more of the enemy's resources. In addition, to the loss of the soldier on the battlefield, resources are required to feed and care for the wounded soldier. However, all this should be considered in the context of the Geneva Conventions (1949) that states it is a war crime to "wilfully cause great suffering, or serious injury to body or health".

The problem of the inability to incapacitate an enemy combatant was first encountered by the British Army in India in the latter part of the 19th C. where full metal jacket ammunition was introduced. That is, it was determined that the Mark II Lee-Metford rifle bullet was insufficient to stop a 'determined rush' and therefore there was a desire to develop a bullet with enhanced stopping power without the loss of velocity and range. The problem was illustrated by a report that a 'tribesman' had been shot six times and after being hospitalised, made a full recovery. Thus, a more venomous bullet was required. This task fell to Captain Bertie-Clay, RA., superintendent of the Dum Dum Ammunition Factory (Penn-Barwell 1896). He developed the soft hollow-point projectile that was to be used in the Lee-Metford rifle. Hollow points were nothing new and had been developed prior to 1896. They were subsequently outlawed in the Hague in 1899 for warfare as being judged to be too inhumane. Again, we can refer to the Geneva Convention (1949) that states that it is a serious violation of law to "[employ] bullets which expand or flatten easily in the human body, such as bullets with a hard envelope which does not entirely cover the core or is pierced with incisions".

The principal reason why these bullets were judged to be inhumane was by the way that the expansion of the bullets caused substantial organ damage and made treatment difficult (due to the number of fragments deposited into the flesh). However, they have some advantage for law enforcement and are still used today for exactly that. Indeed, despite opposition, the New York Police Department introduced such projectiles in 1998. There are several benefits to using this type of projectile in the context of law enforcement. Firstly, it is argued that hollow point projectiles are more likely to come to rest inside the person who is targeted and therefore there is less risk of the bullet leaving the body and causing injury or death to other people. This is a huge problem for law enforcement—particularly in environments where there are hard surfaces to cause ricochet of exiting bullet fragments. Secondly, there is an increased probability that the person who is shot would be severely and rapidly incapacitated. This would be due to the extensive injury that the person would be subjected to and often is essential in hostage situations. Thirdly, it has been argued, soft body armour materials, such as Kevlar® are more effective at stopping soft or hollow-pointed bullets compared to their full-metal-jacket equivalent. Indeed, as we saw in the previous chapter the jacket of the bullet will provide inertial confinement to the projectile core. If the jacket is easily removed due to failure that originates from the void at the tip of the jacket, then the inertial confinement is lost and the soft core is rapidly exposed to the armour. The advantage to law enforcement here is that there would be less risk of 'death by friendly fire'.

Another variant is the soft-point bullet. Soft point projectiles differ from the hollow point in that there is a portion of lead core that is exposed at the core. The effects are similar to hollow points although they are less likely to expand as quickly on contact. It was the soft-point ammunition that was used by Australian law enforcement when they stormed the Lindt Café during the Sydney siege in 2014; a decision that proved deadly to one of the hostages—Katrina Dawson. Although the M4 Carbines have been used extensively in Australia on operations since the 1990s, Ms Dawson was struck by a fragment from a police bullet, despite the use of soft-pointed ammunition (Kidd 2016). Nevertheless, there were contributing factors, namely the environment in which the hostages were held. The Café was previously a bank with hard marble surfaces which may have contributed to a ricochet of a fragment. Further, the police had been criticised for the number of shots fired (22 in all) and the choice of their weapons. Notably, the formal inquest concluded that no criticism was warranted in that regard (Wales 2017). There was a question regarding whether a 'tactical bonded bullet' would have been a better choice of projectile used in the incident. Bonded bullets have the advantage that the lead used in the core of the projectile is physically attached to the jacket by effectively soldering the lead core to the jacket. This is commonly achieved by melting the lead core in situ within a jacket preform to provide a physical bond between the lead and the brass jacket once the lead has solidified. Flux is added to help in the process. This is very similar to the soldering process that occurs when printed circuit boards are assembled. The physical bond between the jacket and the lead core means there is less risk for fragmentation during penetration. However, often it is the lead component of the core that is the weak material that is used in construction and it is unclear whether the use of a bonded bullet would have resulted in a different outcome in the Lindt siege. The use of a lower-velocity firearm may have.

## 8.5  Tumbling Bullets (During Penetration)

It is well established that bullets will tumble during penetration of the human body. A tumbling bullet will, of course, cause more injury than a bullet that is moving in a straight line. So, tumbling bullets are bad. Notably, relatively short bullets (such as the AK 47 projectile) will tumble much later in the penetration path than longer bullets (such as the 7.62 mm NATO bullet). In fact, an AK 47's bullet is not likely to tumble until it has travelled 27 cm or so in fatty tissue. If there was ever a good reason to shed a few pounds before going into battle—this is it!

Of course, bullets do not tumble in flight. They tend to yaw a little as they precess towards the target but usually find stability after 10 m from the muzzle. The cause of the tumbling in the body originates from the different densities and strength of tissue that is penetrated. This causes an asymmetric lateral force to act on the bullet and as the effects of inertia diminishes (as it slows down), the bullet becomes more susceptible to these forces. Clearly, tumbling bullets can and will be detrimental to a person's health. Best to avoid.

## 8.6  How to Penetrate Armour

Armour is intended to be as effective and as lightweight as possible, and not too bulky! Most armour systems are 'passive'. That is, they are simply 'dumb' pieces of material that can defeat the incoming projectile by their inherent mechanical properties (hardness, strength and so on). The choice of materials is important and generally it is best to use materials of low density—particularly when somebody must carry the armour! However, that is not a universal rule. High density materials such as tungsten carbide, tungsten and depleted uranium have all been used in specific armour scenarios.

Armour materials can be divided into two different categories that depend on the way in which they 'manage' the energy of the projectile. Armour materials tend to be either energy 'disruptive' in nature or energy 'absorbing'. Disruptors (or 'disturbers') tend to be made from high-strength materials such as high-strength steel or ceramic. The purpose of these high-strength materials is to blunt the incoming bullet or rapidly erode it. If the projectile is fragmented, a hard material will tend to disperse the fragments and therefore the kinetic energy. An absorber on the other hand works to absorb the kinetic energy of the projectile or fragments through deforming such that the material is stretched beyond its elastic limit and behaves plastically. Through the plastic deformation processes the kinetic energy of the projectile will be converted to a lower form of energy, such as heat. Therefore, the absorbing components of armour systems are generally materials that can undergo large amounts of plastic deformation before they fail. Usually we find that both disruptors and absorbers are designed to work in union—especially in the case of brittle disruptor materials such as ceramic.

The core of the bullet that penetrates armour needs to be much stronger and harder than the core that is intended for 'people penetration'. For projectiles that are designed to penetrate people, the purpose is to transfer the kinetic energy of the bullet into the person rapidly. To achieve this, the projectile needs to be arrested very quickly and this is best achieved by using soft materials in the bullet's construction. These so-called 'ball' projectiles that we discussed in Chap. 7, tend to have cores made from lead-based alloys. As we have seen, lead is soft and malleable and will deform rapidly on contact with rigid structures in the body such as bone. The bullet expands quickly on contact with bone and therefore is subsequently prone to deceleration. However, in many cases this does not happen and the projectile can leave the body fully intact. On the other hand, armour-piercing bullets comprise of cores that are made from materials such as high-strength steel or tungsten carbide. These are hard and strong materials. Tungsten carbide is a wear-resistant, strong material that has been extensively used in cutting tools, screw-driver tips and the like. Due to its high toughness and strength it makes an excellent candidate for armour-piercing projectiles. Pure tungsten is rarely used, generally because of the cost. Tungsten projectiles are generally reserved for specialist weapons such as the areal denial weapons that attack incoming missiles, for example.

So what material properties are best for an armour design?

The golden rule with armour design is its ability to absorb energy or redirect the energy of the projectile. So, toughness is an excellent property to have. Strength too is an important property in that regard. Many armour materials rely on the strength to smash the incoming projectile. Hardness is another way to describe the strength of the material. Hardness is measured by pressing a diamond indenter into the material and quite simply, the bigger the indentation, the softer the material. So, hardness is an important factor, in relation to metals and ceramics that is. As a general rule of thumb, the harder the material, the better the performer. Or so we thought! The ceramic boron carbide is an excellent potential armour material due to its insanely high hardness, being close to half the hardness of diamond and low density. Yet at high impact velocities it is softened by the extensive damage that occurs in its microstructure. In fact, as the impact velocity is increased, researchers have noted that it gets softer. Of course, relatively speaking—I am talking about the hardness of a common-grade aluminium oxide, which would still break your teeth if you bite into it!

The way that materials respond to high stress is still subject of intense scientific curiosity. And this is all the more important when we consider the design of armour systems. Armour can fail through a number of complex processes that originate at the atomic level, and these can range from plastic flow that results in stretching of the material and subsequent petalling,[3] right through to tensile spall failure, where the material is ripped apart by shock wave reflections.

Materials fail in a complex fashion and there is still a lot we do not understand about plasticity mechanisms, failure initiation and propagation. It is fair to say that every year around the world there are numerous conferences for academics working on these problems. They are trying to understand how materials fail under specific loading conditions, some of which are quite extreme. For example, there is a whole conference organization system devoted to the study of ballistics and organised by the International Ballistics Committee. This is a conference that meets every 18 months and there will be numerous people from various government institutions, academic institutions and companies who meet together to discuss and learn about how materials fail under impact loading. Other conferences include DYMAT (Dynamic Behaviour of Materials), conferences organised by The Minerals, Metals and Materials Society as well as the conferences organised by the American Physical Society, such as the Shock Compression of Condensed Matter series of conferences.

---

[3]So described, as this failure mode looks like a blossoming flower.

There are probably a few thousand papers published annually on how materials and structures behave under dynamic stimuli and yet there is so much more to learn. Ever since the turn of the 19th C. and the likes of Jean-Victor Poncelet (1788–1867) who published a very simple model of projectile penetration (Backofen Jr 1980) have researchers been eager to share their work. Despite the 'hush, hush' nature of defence projects, there is an awful lot of defence-relevant information that is actively published and pushed into the public domain. And, for good reason. It is the best way that knowledge can be retained, without the risk of it disappearing into a dusty filing cabinet in Defence HQ.

## 8.7   Soft Armour

Most materials used for soft armour that can be easily warn by an individual are fibrous in nature. That is to say that they consist of multiple plies (or layers) of fibres. These tend to be used in body armour systems with the most famous of the fibres being Kevlar® which is mostly used in a woven fabric construction. Kevlar® was first developed by Du Pont in the 1960s from aromatic polyamides. The discovery of these fibres occurred purely by accident. Stephanie Kwolek (1923–2014), was a chemist working at DuPont and she, along with her colleagues, were trying to develop a replacement to steel fibres used within road tyres. Engineers believed that replacing the relatively high-density steel fibres with a low-density polymer fibre would lead to improvements in fuel efficiency. The discovery of what became the Kevlar® fibre was a stunning achievement. It had excellent strength and stiffness properties and was several times stronger than steel on a weight-for-weight basis. This, and similar fibres such as Twaron® manufactured by Teijin Aramid, are known as aramids and are derived from polymer molecules containing benzene rings. These molecules readily align parallel to each other to form highly ordered structures and consequently demonstrate excellent properties along the direction of the fibres.

Of course, it should be of little surprise that fibrous composites are used in protection systems given the very early use of leather as armour. Leather is fibrous in nature with the outer layer (the grain) consisting of the epidermis of the animal whereas the corium underneath, provides a loosed fibrous structure. The structure is tough and consequently it has been used for hundreds of years dating back to before the Qin dynasty of Ancient China (ca. 221 BC). Arguably, wood (another fibrous composite) has been known for millennia for its protective properties and in particular, its ability to absorb the energy of a low velocity impact. We still use it for that today.

Generally, there are three principal properties in a fibre that are required maximising protection whilst minimising discomfort for the wearer. The first is toughness (or energy absorbing ability). If we were able to pull a fibre so that it stretches and in doing so were able to measure the force acting on the fibre we would be able to plot the force with the displacement. The area under the curve[4] would represent the energy absorption capacity. This is a measure of toughness. When the curve is plotted in terms of stress and strain, then the area under the curve would be the energy absorption capacity *per unit volume*. A high elastic wave velocity is also important as this allows for the rapid delocalization of the stresses from the point of impact. So as the projectile collides with the fibre the energy is dissipated from the point of impact rapidly. Finally, a low density is important. And this is why materials such as Kevlar® and ultra-high-molecular-weight-polyethylene (UHMWPE) such as Dyneema® are so useful in armour constructions. These materials have densities close to water. With UHMWPE, the density is less than water so it will tend to float!

One such fibre that has all this potential and has already been used in body armour is PBO (poly p-phenylene-2,6-benzobisoxazole) currently manufactured under the trade name Zylon®. This fibre has a tensile strength of 5.2 GPa (many more times that of a high strength steel) and all the mechanical metrics suggested that it would perform as an excellent ballistic-grade material. However, a report from the National Institute of Justice (US) suggested that it degrades due to environmental conditions (moisture and heat (Walsh et al. 2006a)) and this, it was thought, was a contributing factor to the failure of a vest warn by a police officer who was mortally wounded. Worryingly still, the report implied that a visual inspection of a Zylon®-based body armour would not indicate whether the intended ballistic performance was maintained (Hart 2005). There has been subsequent attempts to stabilize PBO fibres but to date these have been deemed as not being successful (Walsh et al. 2006b). It is no longer used in body armour.

## 8.8 Hard Armour

Probably one of the more unusual choices for armour is that of ceramic. Most people associate the word ceramic with sanitary ware, tiles and pottery which are silicate-based ceramics. These are quite porous, non-uniform in microstructure and quite weak. Whereas from a technical point of view, the ceramics that are of interest for armour applications need to be strong and resilient to impact. These are not silicate-based materials and depend on much more sophisticated raw materials such as oxides or carbides. With these materials, microstructures can be produced that are finer, and consistent throughout the thickness of the material. They tend to have little-or-no porosity which are small holes in the material's structure from which cracks can appear. These materials are often referred to as *engineering ceramics* or *advanced ceramics*.

It has been known for over 100 years that bonding very hard materials to metals can improve their ballistic performance. Maj Neville Monroe-Hopkins found that a thin layer of enamel improved the ballistic performance of a thin steel plate (Dunstan and

---

[4]Rarely would the force-displacement curve appear as a curve, in fact most fibres exhibit a linear response all the way to failure.

Volstad 1984). This work was carried out in 1918. Enamel is a fused glass. It is made by fusing powdered glass to a substrate at temperatures of between 750 °C and 850 °C to provide a thin layer of a hard glassy material. This is commonly found on metal sinks, baths, cups, pots and so on. Technically, I would argue that Monroe-Hopkins was the inventor of ceramic-faced armour as glass is regarded as a part of the family of ceramic materials, which also includes clay products, refractories, abrasives and cements (Callister and Rethwisch 2009). Irrespective, he clearly developed one of the principles of ceramic-armour design, namely: placing a hard, brittle structure onto a relatively ductile backing layer provides good ballistic protection.

It was not until the 1960s however that patents started appearing for ceramic-faced armour. Patents were filed by the Goodyear Aerospace Company, with the first filed in 1963 and granted in 1970 (Cook 1970; Cook et al. 1979). The initial patent application detailed a ceramic-faced armour comprised of a ceramic facing attached to a woven fabric composite substrate. The invention was timely: the first real use of ceramic-armour technology was in US helicopters during the Vietnam conflict where low-level sorties made the helicopter and crew vulnerable to small-arms fire. In 1965 the first ceramic-based aircrew body armour vest was manufactured as this was the most weight-efficient means of providing protection (Bart and Lindberg 1987). Also, in 1965, the UH-I 'Huey' was fitted with a 'Hard Faced Composite' (HFC) armour kit used in the armoured seats for the pilot and co-pilot. The seats provided protection against 7.62-mm AP ammunition on the seat bottom, sides, and back due to the use of a boron carbide face and fibreglass backing. In 1966, the first monolithic ceramic body armour vest was issued to the helicopter crews along with other protection improvements including the use of airframe-mounted armour panels. It has been estimated that, between 1968 and 1970, these improvements in aircrew armour reduced the number of non-fatal wounds by 27% and fatalities by 53%. Ceramic armour had proved its usefulness.

## 8.9   Complex Armour

Defeating small arms bullets is quite a straight-forward process if you know what to do and what to look for. Arguably it is more challenging to stop an armour piercing round where the core would be a high density, hard material and it would be travelling at a high velocity. However, probably the most troubling challenge is to defeat an incoming Armour-Piercing Fin-Stabilised Discarding Sabot (APFSDS) projectile, the type that is fired from a tank gun. The way that this projectile is defeated is by using multiple layers of strong materials. Some examples of these types of armours have been code-named 'Chobham' and 'Dorchester'. In fact, there is still quite a substantial amount of mystique around these types of armours. Consequently, their recipe is not publicised and kept secret—for obvious reasons.

The mechanism by which these types of projectiles are defeated is by a process of material erosion. As we have seen previously, these projectiles appear to penetrate the target hydrodynamically. This means that the interaction between the projectile and the target behaves as if it is if they are both acting like fluids. It is known that ceramics in particular are resilient against these types of threats mainly because they maintain very good strength properties under very high velocities and high pressures. Therefore, we would do well to assume that the 'mystical' complex laminate armours comprise of some recipes that would involve the use of a hard, strong material such as a ceramic. It is also well publicised that certain M1 Abrams main battle tanks come with depleted uranium that is used in the turrets. Again, the reason for this is not well publicised.

The other challenge for heavy armour is that vehicles must be designed to take into account hits by a range of different munitions. These may include improvised explosive devices, explosively formed projectiles, shaped charge jets and so on. This presents a challenge in maximising protection whilst minimising weight. Therefore, we have to come up with alternative approaches to defeat a range of threats. Other suggestions for defeating APFSDS projectiles and shaped-charge type systems include the use of active defence systems. These are systems that detect, track and launch an effector to deal with the incoming projectile. This is especially challenging as certain APFSDS projectiles can be travelling at a velocity in excess of 1500 m/s and therefore there has to be a quick reaction system on board the vehicle. Furthermore, these types of projectiles are particularly nasty in that they have substantial inertia. That is, they have a high mass and high velocity. This means that it is difficult to knock such a projectile off course, and indeed destroy it. Nevertheless, there are certain things that you can do by focussing on the fins of the projectile (recall that these types of projectiles are drag stabilised). The fins of the projectile are very often the Achilles' heel to the projectile's survivability from an effector. If the fins can be destroyed, or even weakened, then the stability of the projectile is substantially compromised (Hazell 2015).

## 8.10 Reactive Armour

So far, we have been talking about the subject of passive armour. However, there are armour systems that literally react to the incoming projectile. The most famous of those is called explosive reactive armour or 'ERA'.

It may seem non-sensical to use explosive is an armour however a little-known fact is that the role of explosives in intercepting the high velocity Munroe jets was established towards the end of World War II. However, this approach was discounted as it was largely thought that surrounding a tank with explosives was not practical or desirable. Other possibilities included the use of oxidising agents such as nitrates and chlorates, dry sand, and water. Based on a limited number of experiments, it was deduced that sodium carbonate decahydrate (washing soda) was the most effective of oxidising agents (Roberts and Ubbelohde 1944). Even 'luting' (liquid clay/cement) was considered and measured to be as high as five times as effective as steel on a weight-by-weight basis (Anon 1944). Interestingly, ceramics and even 'explosively-driven' ceramics have also since been discovered as being good at destroying shaped charge jets (Hazell et al. 2012).

Like all good scientific discoveries, ERA a was discovered by accident. This was invented by a German scientist called Manfred Held (1933–2011) back in the late 1960s. After the 6-Day War in 1967, Held travelled to the Middle East to test the Soviet tank armour with shaped-charge warheads. He noticed that on testing the turrets on some of the tanks that the on-board ammunition detonated and somehow cancelled the blast of the anti-tank shaped-charge warhead. But why? The simplest explanation was that the momentum of the rapidly expanding blast wave from the munitions was disrupting the jet. That conclusion led Held to develop explosive reactive armour.

The simplest construction of ERA consists of two steel plates sandwiching a layer of high explosive. When the jet penetrates the outer housing and the first steel plate it rapidly compresses the high explosive. The rapid compression of the explosive leads to detonation, propelling the steel sandwich plates apart. Frequently, the leading edge of the jet perforates through both plates before the flyer-plate starts to move and therefore escapes any interaction. This portion of the jet is called the precursor and its length is somewhat determined by the obliquity of the cassette and the velocity of the jet. To maximise protection, it is necessary to accelerate the plates to a very high velocity to maximise the amount of material offered to the incoming shaped-charge jet. The plate velocity depends on its mass and the type and mass of the explosive. Finally, for optimum disruption, the ERA cassette is angled to the incoming threat. Therefore, the outer steel plate moves across the path of the jet thereby continually offering fresh steel to perforate—cutting a slot in the moving plate or plates. Thus, the steel plates from the armour are literally 'unzipped' by the incoming jet and the resulting penetration is called the 'zipper effect'.

# References

Anon (1944) The protection of steel targets from attack by shaped charges. Ministry of Supply, London, UK

Backofen Jr JE (1980) Armor/ Armor penetration: land, sea, air and space. In the Proceedings of the 5th International Symposium on Ballistics, Toulouse, France, 16–18 April 1980

Bart RK, Lindberg JC (1987) Ceramic bodyguards. Adv Mater Process 132(3):69–72

Blanchfield P (2015) The sometimes tragic physics of celebratory new year's gunshots. The Trace. https://www.thetrace.org/2015/12/celebratory-gunfire-new-years/. Accessed 10/11 2020

Brown AD, Walters JB, Zhang YX, Saadatfar M, Escobedo-Diaz JP, Hazell PJ (2019) The mechanical response of commercially available bone simulants for quasi-static and dynamic loading. J Mech Behav Biomed Mater 90:404–416. https://doi.org/10.1016/j.jmbbm.2018.10.032

Callister WD, Rethwisch DG (2009) Materials science and engineering: an introduction, 9th edn. Wiley, New York, US

Cook RL (1970) Hard faced ceramic and plastic armor. US Patent

Cook RL, Hampshire WJ, Kolarik RV (1979) Ballistic armor system. US Patent

Dunstan S, Volstad R (1984) Flak jackets: 20th century military body armour. Osprey Publishing Ltd., London

Fackler ML (1988) Wound ballistics. a review of common misconceptions. J Amer Med Assoc 259(18):2730–2736

Fackler ML, Dougherty PJ (1991) Theodor Kocher and the scientific foundation of wound ballistics. Surgery Gynecol Obstetrics 172(2):153–160

Fackler ML, Surinchak JS, Malinowski JA, Bowen RE (1984) Wounding potential of the Russian AK-74 assault rifle. J Trauma 24(3):263–266

Haag LC (1995) Falling bullets: terminal velocities and penetration studies. Wound Ballistics Rev 2:21–26

Hanieh A (1971) Brain injury from a spent bullet descending vertically. Report of five cases. J Neurosurgery 34(2 Pt 1):222–224. https://doi.org/10.3171/jns.1971.34.2part1.0222

Hart SV (2005) NIJ Special Report: third status report to the attorney general on body armor safety initiative testing and activities. U. S. Department of Justice, Office of Justice Programs, National Institute of Justice, Washington, DC, USA

Harvey EN, Korr IM, Oster G, McMillen JH (1947) Secondary damage in wounding due to pressure changes accompanying the passage of high velocity missiles. Surgery 21(2):218–239. Export Date 26 May 2013

Hazell PJ (2015) Armour: materials, theory, and design, 1st edn. CRC Press, Boca Raton

Hazell PJ, Lawrence T, Stennett C (2012) The defeat of shaped charge jets by explosively driven ceramic and glass plates. Int J Appl Ceramic Technol 9(2):382–392

Hollerman JJ, Fackler ML, Coldwell DM, Ben-Menachem Y (1990) Gunshot wounds: 1. Bullets, ballistics, and mechanisms of injury. Amer J Roentgenol 155(4):685–690

Kidd J (2016) Sydney siege inquest: NSW police commander defends M4 rifle and ammo used by tactical officers. ABC News. Accessed 29 April 2019

La Garde LA (1916) Gunshot Injuries: how they are inflicted, their complications and treatment, 2nd edn. William Wood and Company, New York

Morgan MM, Spencer AD, Hershey FB (1961) Debridement of civilian gunshot wounds of soft tissue. J Trauma Acute Care Surgery 1(4):354–360

Ordog GJ, Dornhoffer P, Ackroyd G, Wasserberger J, Bishop M, Shoemaker W, Balasubramanium S (1994) Spent bullets and their injuries: the result of firing weapons into the sky. J Trauma—Injury Infection Critical Care 37(6):1003–1006. https://doi.org/10.1097/00005373-199412000-00023

Penn-Barwell J (1896) The military bullet. British Med J (2)

Roberts CE, Ubbelohde AR (1944) Protection of steel targets by non-metallic substances with special references to oxidising agents. Target Damage by Munroe Jets. Armament Research Department

The New York Times (1898) In: Science and industry. The New York Times. 27 November 1898

Wales SCoNS, (2017) Inquest into the deaths arising from the Lindt Cafe siege: findings and recommendations. NSW Government, Sydney, Australia

Walsh PJ, Hu X, Cunniff P, Lesser AJ (2006a) Environmental effects on poly-p-phenylenebenzobisoxazole fibers. I. Mechanisms of degradation. J Appl Polym Sci 102(4):3517–3525. https://doi.org/10.1002/app.24788

Walsh PJ, Hu X, Cunniff P, Lesser AJ (2006b) Environmental effects on poly-p-phenylenebenzobisoxazole fibers. II. Attempts at stabilization. J Appl Polym Sci 102(4):3819–3829. https://doi.org/10.1002/app.24794

Young L, Rule GT, Bocchieri RT, Walilko TJ, Burns JM, Ling G (2015) When physics meets biology: low and high-velocity penetration, blunt impact, and blast injuries to the brain. Front Neurol 6:89–89. https://doi.org/10.3389/fneur.2015.00089

# Chapter 9
# Gun Ownership and Gun Control

As we have seen, it is reckoned that there are close to 1 billion firearms in existence in the world today and 85% of these guns are owned by civilians. In this chapter we are going to examine gun ownership and whether this is in fact harmful to the individual and society as a whole. Guns do serve a purpose. They are not only used for maintaining law and order by police, but citizens often possess firearms for self-protection, recreational use and in rural settings, for dealing with pests (of the animal variety).

Many people have a well-considered position on this subject and my intention is not to offend. Debate rages over the benefits of gun control measures and no more so than in the United States which has the highest level of gun ownership per capita than anywhere else in the world. Even scientists hold differing views on this subject and appear to interpret the raw data differently. When reading the academic literature it is sometimes difficult to distinguish advocacy from fact. This is not helped by the nature of social science research which is generally plagued by a reproducibility crisis. That is, scientists struggle, in some cases, to reproduce the findings of their peers. In one study published in 2018, nearly 40% of social science research published in the premier journals of *Science* and *Nature* could not be reproduced (Camerer et al. 2018). We are not talking about common magazines here, we are talking about two of the most prestigious journals for peer-reviewed research on the planet! The papers covered in this article would have been thoroughly reviewed by some of the top scientists in the world. This is deeply worrying for the social sciences as it is for wider society. It also begs the question, has anyone tried to replicate this reproducibility study? Anyway, I digress. There are people on both sides of the argument that hold well-meaning views and in this chapter, we are going to try and address these, and where possible, see what the science has to say.

Different nations have different approaches to gun control and these approaches have come about by addressing the balance between the rights of an individual to protect their family's property and society's need for maintaining law and order and reducing overall firearms deaths. And these two requirements (if we can call them 'requirements') appear to be working as opposing forces in some cultures.

© Springer Nature Switzerland AG 2021
P. J. Hazell, *The Story of the Gun*, Springer Praxis Books,
https://doi.org/10.1007/978-3-030-73652-1_9

There has been mass shooting in many countries around the world. However, the US *appears* to win the prize for the most prolific numbers of shootings where the target has been random strangers (although we will re-examine the evidence on this again in a moment). Nowhere is this better epitomised than in the case of Stephen Craig Paddock. In 2017, Paddock, a 64-year-old retired accountant with no previous criminal record, set himself up in a hotel room of the 32nd floor and fired into a crowd of 22,000 (random strangers) who were attending the Route 91 Harvest Festival in Las Vegas. The density of the crowd and his vantage point clearly had a bearing on the number of casualties. Fifty-eight people lost their lives. Police say that they found 23 guns in his hotel room; Paddock must have brought them into his hotel room in bags—probably making eight or so round trips. The calibres that were found in Paddock's room included 0.308-inch / 7.62 mm, and 0.223-inch / 5.56 mm. All semi-automatic weapons used in the shooting were obtained legally. His actions were clearly deliberate.

This was America's most deadly mass shooting and trumped the mass shooting that happened the previous year at the Orlando Pulse night club in June 2016 where 49 people lost their lives. Again, Omar Mateen, the shooter, obtained his guns legally. In fact, it turned out that a friend of our family was present with the first-responders in the aftermath. So, the question that must be asked and has been asked countless times by journalists, commentators and politicians, is: does gun control save lives? If there was an increased level of gun control in the US, would that have saved the lives of the 58 teachers, military veterans, nurses and teenagers that Paddock took?

Or take the case of Robert Bowers, 46, who opened fire at the Tree of Life synagogue during its Sabbath service where a baby naming service was occurring. Eleven worshippers lost their lives. Armed with an assault rifle and three handguns that he legally owned he emptied his magazines into unsuspecting victims in what appeared to be an act of anti-Semitic hate.

And there are many more cases of where deranged killers decide to go on to take the lives of people that they do not know. Random strangers that are in the wrong place at the wrong time. A spiteful act of violence, where men, women and children are massacred for no reason other than to leave this Earth in an act of unparalleled violence.

By the time you come to read this book you would have probably forgotten the horror of Paddock's attack, or that of Bowers and there would have probably been more mass shootings. And, not just in the US. Mass shootings in the US have become notoriously common on our TV screens and we become fascinated with the subject of gun control for a short period of time after a mass shooting. However sadly, there is a collective loss of memory as time goes by until the next set of homicides and the politicians appear powerless to make the necessary changes to legislation. Notably, we see that Google searches for 'gun control' spike after a mass shooting but diminish rapidly as time goes by. Society has a collective amnesia over these things.

**Table 9.1** Firearm and non-firearm homicides in the frontier countries (per 100,000, 1999–2000); after (Hemenway 2017)

| Nation | Firearm homicide rate | Non firearm homicide rate | Total homicide rate | Households with guns (%) |
|---|---|---|---|---|
| USA | 4.0 | 2.2 | 6.1 | 41 |
| Canada | 0.6 | 1.2 | 1.8 | 26 |
| Australia | 0.4 | 1.4 | 1.8 | 16 |
| New Zealand (1997-1998) | 0.2 | 1.5 | 1.7 | 20 |

## 9.1   The Scope of the Gun Problem in the USA

Worldwide it is estimated that 251,000 people lost their lives to firearm injuries in 2016. Six countries (Brazil, United States, Mexico, Colombia, Venezuela, and Guatemala) accounted for 50.5% of the Global death toll (Nand et al. 2018). Out of all the developed English-speaking nations, the USA does have a gun-related crime problem. This is beyond doubt. When we look at a broad range of crime committed (car theft, burglary, robbery, sexual violence and so on), countries such as the USA, Australia, New Zealand and Canada (i.e., the so-called 'frontier countries') have broadly similar crime levels. If anything, it appears that the USA has the lowest crime rate[1] per capita when compared to these other nations when looking at crimes that do not directly lead to a homicide. So, that means the prevalence of guns results in less overall crime, right? Maybe.[2] However, what distinguishes the US from these other countries is the relatively high rate of homicides and how many are committed with guns. This latter fact appears to correlate with gun ownership (see Table 9.1).

So, it seems that in the USA, most of the homicides are committed with firearms whereas in the other frontier countries the opposite is true. Furthermore, when we compare violent deaths of five to fourteen-year-olds with other twenty-five high-income countries during the 1990s we see that the US fairs very badly. It has been shown that the US had a higher suicide and homicide rate due to the use of guns and a nine-times higher rate of unintentional gun deaths when compared to other wealthy nations (Hemenway 2017).

### 9.1.1   The Second Amendment

In the United States of America most of the arguments supporting the right to bear arms come from the second amendment of the constitution. It is clear that the

---

[1]The term 'rate' here is indicative of the number of incidents in the population over a period of time (e.g., 1 year). We will see this used again when we discuss suicide rates and so on.

[2]There are many contributing factors that lead to a reduced crime rate including, approach to law enforcement, political and socio-economic factors, levels of inequality and so-on.

'founding fathers' were committed to protecting the rights of the individual being trodden down by an overbearing government and the notion of flourishing under freedom is the focus of the Constitution and the ten amendments that became the Bill of Rights.

Much of the focus of the gun control argument in the US gravitates on The Second Amendment which reads as follows:

> A well-regulated Militia being necessary to the security of a free state, the right of the people to keep and bear arms shall not be infringed.

Now of course, the US was a very different place when The Second Amendment was ratified in 1791. It had not been long since the part of the Americas had fought off British colonisation and it was clear that an armed militia needed to be established. The notion of a militia is also mentioned in the Fifth Amendment. The purpose of a militia was to protect each state from threats outside (e.g., invasion) and threats from within (e.g., riots). Many Americans would believe that the Second Amendment gives them the right to bear arms for the purposes of self-defence and to allow them to oppose tyranny. However even though the historical arguments on the Second Amendment are somewhat 'muddy', there does not appear to be too much intellectual support for an individual's right to bear firearms, at least by virtue of the Second Amendment alone. For a detailed discussion on this, I highly recommend the reader to see Chap. 8 of Hemenway (2017).

It can be argued that when governments in the past have moved to confiscate firearms, gun control was a step towards genocide. And this appeared to be true, at least, for Hitler and Stalin. It has been claimed that gun registration lists were used to confiscate guns, after which the government could be allowed a free-path to genocide. And this, it has been claimed (Kopel 2013), has been the pathway in *every* episode of genocide in the past century including: Turkish Armenia, the Holocaust, the USSR, Soviet-occupied Poland, Guatemala, Mao's China, Chiang Kai-Shek's White Terror, Uganda, Cambodia, Srebrenica, Zimbabwe, Darfur and so on.

So, should we be fearful of an impending genocide if our government wants to impose gun control? Other warning signs also preceded many, if not all of the above-mentioned genocides. These included citizen registration processes (yet nobody would argue against passport ownership today and many countries have adopted ID cards without the threat of an impending genocide.).

What is perhaps remarkable is that this is frequently used as an argument in the context of the US where the military is one of the most powerful in the world. Yes—it is true that if one has the right to bear arms then one can protect one's family from thugs but to think that through the ownership of pistols, rifles and assault rifles that you can fight against a well-armed (and well armoured) government is somewhat 'optimistic' and bordering on fantasy. Let's put this in context further: if I wanted to shoot a hole through an M1 Abrams Battle tank I would need a very powerful shaped charge munition or a gun with *at least* a 105 mm calibre and capable of firing a heavy-metal armour piercing fin stabilised projectile. Such weaponry is usually out of the reach of even the most ardent NRA[3] supporter.

This argument is perhaps one of the more tenuous arguments against gun control. All Western Governments these days maintain a highly-trained, highly-equipped armed force and it is illogical to suggest that one individual or even a militia armed with semi-automatic weapons can put up any kind of resistance to their government. The only way a militia would stand any chance to militarily bringing down a government is by using weapons of mass destruction—a thought that most decent people would find abhorrent.

However, there appears to be little doubt that where faith in the government is lost in terms of the protection that it can offer to one's family then gun ownership increases—and in some cases understandably so. Take the town of Michocan in Mexico—the subject of the documentary "Cartel Land". It is no surprise that people wanted take up arms against the horrors that were unleashed by the drug lords. The people had had enough! And, so the Autodefensas was born—a well-armed vigilante group. Mexico is perhaps unusual. However, the right to bear arms in this case, is certainly viable when law enforcement fails to protect the vulnerable.

## 9.2  Mass Shootings

Why is it that people who have, it seems, an open access to weapons carry out such insane acts of violence on innocent victims? Why is it that apparently normal people who have jobs, families and normal hobbies feel obliged to use weapons to take so many lives?

Does the US lead the world in the number and lethality of mass shootings? From the television news that we receive on a daily basis, one would assume yes. And in fact, according to a 2016 academic study, the answer would still be "yes". (Lankford 2016). However, it all depends on what you would define as a mass shooting. A commonly used definition is used to define a mass murder, which in turn has been copied to define a mass shooting as outlined by:

> "a multiple homicide incident in which four or more victims are murdered—not including the offender(s)—within one event, and in one or more geographical locations relatively near one another. It follows then that a "mass shooting" could be defined as a multiple homicide incident in which four or more victims are murdered with firearms—not including the offender(s)—within one event, and in one or more locations relatively near one another." (Krouse and Richardson 2015; Douglas et al. 2006)

It seems that potentially this could be a rather odd definition because if a shooter shot 40 people and only 3 died as a result of their gunshot wounds then it would appear that this would not be classed as a 'mass shooting'. Most people would argue otherwise but this appears to be a widely accepted definition.

---

[3]NRA = National Rifle Association.

So, does the US lead the world in mass shootings? According to Lott (2018), no! In fact, it turns out that when you look at all of the world's data, which admittedly in some places is hard to confirm, the US rate is actually lower than the global average. Lott looked at 1491 cases from 1998 to 2012 and even though in this time the US had approximately 4.6% of the world's population, it had 2.88% of the mass public shootings.

OK, so is there any correlation with civilian gun ownership and mass shootings? After all, the US has by far the highest take up in civilian gun ownership. Well if we take Lott's data and compare that with data from the Small Arms Survey (Karp 2018), then it would appear not. Figure 9.1 shows a plot from the analysis done by Lott and correlated with data from the Small Arms Survey. One would expect to find a clear linear trend if there was an indeed a correlation between gun ownership and mass shootings. There is none.

Critics of this simplistic graph would rightly point out that many of the countries that have elevated numbers of mass shootings are countries that have struggled to curb gang violence or have been at war. Norway is an unusual addition and arguably a blip on the landscape in what is a country that has historically avoided gun violence. This was solely down to the case of Anders Behring Breivak, a far-right terrorist, who killed 69 people at a youth league event in Utoya, Norway in July 2011. However, there are other European countries including Russia, Switzerland, and Finland that all score worse than the US in mass shootings. The question the currently eludes answer, however, is whether the US leads the world where mass shootings have been

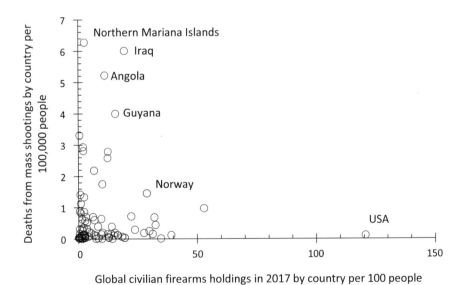

**Fig. 9.1** Deaths per mass shooting by country per 100,000 of the population as a function of civilian firearms holdings. Data taken from (Lott 2018) and (Karp 2018). Cases of death recorded from 1998–2012

perpetrated by a disturbed individual with no apparent ideological or criminal agenda apart from taking life for the sake of it? The media coverage would lead you to think so. However, as we know not all countries are as media-soaked as the US and this impression is probably due to our oversaturation of 24/7 coverage that comes from Western democracies and the continued coverage and commentary on such incidents. This is still an open question.

### 9.2.1 Mental Health Issues

In 2013 a Gallup poll asked the question of US adult residents:

> Thinking about mass shootings that have occurred in the United States in recent years, from what you know or have read, how much do you think each of the following factors is to blame for the shootings—a great deal, a fair amount, or not at all? (Saad 2013)

A list of options was then provided. Of the 1023 adults interviewed, 48% of the respondents thought that failure of the mental health system to identify individuals who are a danger to others was the most important reason for mass shootings. This sat above easy access to guns (40%), drug use (37%), violence in media (32%), the spread of extremist views (29%), insufficient security (29%) and, inflammatory language from political commentators (18%).

So, the public perceive that mental health may be a mitigating factor in mass shootings and they may be right. The mental health of gun users is one of the areas of common ground perpetrated by the gun lobbyists and the gun control advocates alike. Both groups would be keen to see some type of restriction to those diagnosed with mental illness. The fact is that it is known that people with serious mental illnesses are somewhat more likely to commit violent acts towards other people than people who are not mentally ill. However, importantly, the large majority of mentally ill individuals are not violent toward others (Swanson et al. 2015). Further, how do you legislate for this whilst avoiding stigmatism of those who suffer from mental health problems? (see (Appelbaum 2013)). Mass shooters are unlikely to pursue treatment due to their tendency to blame others for their circumstance and consider themselves as victims. Mass murderers see the problem to reside in others, not themselves. Further, no clear relationship between psychiatric diagnosis and mass murder has ever been established (Fox and DeLateur 2014). It is a tricky problem to solve and one that appears to not have a clear answer.

### 9.2.2 Bump Stocks

Mass shootings are more tragic where the shooter uses what is called a 'bump stock'. For example, Stephen Paddock used such a device at the Las Vegas country music festival in October 2017. Yet it still took 443 days for the federal government to ban the attachment.

Bump stocks are devices that convert a semi-automatic rifle to automatic fire. With a semi-automatic weapon, the cartridge is automatically loaded into the chamber by recycling the propellant gasses and firing occurs by pulling the trigger. With a bump stock, firing occurs by literally 'bumping' the trigger against the shooter's trigger finger and thus automatic fire occurs. This is accomplished quite simply by replacing the stock of the semi-automatic weapon with one that allows for spring-actuated linear movement of the trigger piece. The firer's finger is kept stationary by the addition of a resting system so that as the trigger assembly moves forward the trigger is pressed against the finger to fire the weapon. Therefore, firing occurs repeatedly.

The results can be quite astonishing and substantially increases the rate of fire for semi-automatic weapons. It allows a shooter to simply 'spray' a target area with bullets. There are three drawbacks for this type of addition. Firstly, the cost of the ammunition. Automatic fire can literally chew through ammunition and ammunition can be expensive. An AR15 bump-stocked rifle would empty a 40-round ammunition clip in a matter of tens of seconds. Secondly, accuracy is severely compromised with automatic fire compared to semi-automatic discharge where there is more control of the gun. For the majority of recreational shooters who take pride in their craft, accuracy is of paramount importance. Thirdly, the modification would substantially increase the risk of overheating. As we have seen, the heat generated from sustained fire can be substantial and for a semi-automatic gun, which may have not been designed to accommodate the heat, this could cause damage to the weapon. With this is mind, the question would have to be asked, why would a recreational shooter want to use a bump stock?

### 9.2.3   The Odd Relationship Between Mass Shootings and Stock Prices

One of the oddities that we see in the US particularly is that after a mass shooting and amongst the condemnation of political leaders and such, is that stock prices actually rise. This is probably driven mostly by the fear that gun laws would be tightened as a direct result of the mass shooting, resulting in a spending spree by people who would be considering buying a gun anyway. We should bear in mind in that many respects, gun ownership is cultural. For example, in the US it is well known that gun sales spike at Christmas. Since 1998, December has been the busiest month in all but two years (2008 and 2013) for gun sales indicating that when in most countries with active gun control policies, parents would buy their children phones, games and toys in the US, guns will be found on 'Santa's list'.

## 9.3  Is Gun Ownership Harmful?

This may seem a silly question. What does the science say about gun control? Does it work or not? What are the main arguments used by proponents? Do they make sense? Let us have a look at a few of the opposing arguments.

### 9.3.1  More People Are Killed by Hands and Feet, Knives, Hammers, Medical Malpractice

This is an argument that regularly appears on my Facebook feed after a mass shooting in the US. And it may shock you to learn that of course, this is true. More people are killed by hands, feet, knives, hammers and medical malpractice. And, as the conservative Blogger and outspoken gun proponent Bill Whittle says "No one talks about limits to hammers, or knives or doctors or hospitals." But why would we? Hammers are not designed and sold with the sole intention to be the tool that is used in a homicide. Hammers are tools that are designed to hit nails into wood. The same could be said about motor vehicles which cause more deaths than all the homicides put together in the US. However, the main purpose of the motor vehicle is to get you from one place to another place usually quite safely. Hospitals are there for the common good and Doctors try their utmost to preserve life rather than take it. The point that I am making here is that none of these things are objects that have a sole purpose of causing injury or killing. The gun on the other hand is. A gun has no other purpose in its design but to destroy, injure or take a life. Even the trusty knife that could be argued as an efficient murder weapon can be bought for honourable purposes. Not least to cook with. So, in my mind this is rather a straw-man argument because *anything* can kill you really.

### 9.3.2  More Guns Reduce Burglary Rates

The evidence does not support this (see Table 9.2). There have been several international studies where victims are surveyed with their experience with crime. The International Crime Victimisation Survey (ICVS) is the most far-reaching programme of fully standardised sample surveys looking at householders' experience of crime in different countries. Looking at the data from 1999[4], it is clear that the USA, which has one of the most relaxed gun control policies in the world, that the rate of burglaries is around average. Attempted burglaries are comparable to England and Wales (which has arguably the strictest gun-control legislation) and sits bang on the international average where burglary with entry is recorded.

---

[4]Studies from other years report similar conclusions.

**Table 9.2** Attempted burglar and burglary with entry in 1999 from the international crime victims survey; % victimised once or more in 1999

| Country | Attempted burglary (%) | Burglary with entry (%) |
|---|---|---|
| Australia | 3.3 | 3.9 |
| Belgium | 2.8 | 2.0 |
| Canada | 2.3 | 2.3 |
| Catalonia (Spain) | 0.6 | 1.3 |
| Denmark | 1.5 | 3.1 |
| England & Wales | 2.8 | 2.8 |
| Finland | 1.0 | 0.3 |
| France | 1.3 | 1.0 |
| Japan | 0.8 | 1.1 |
| Netherlands | 2.7 | 1.9 |
| Northern Ireland | 0.9 | 1.7 |
| Poland | 1.3 | 2.0 |
| Portugal | 1.2 | 1.4 |
| Scotland | 1.9 | 1.5 |
| Sweden | 0.7 | 1.7 |
| Switzerland | 1.8 | 1.1 |
| USA | 2.7 | 1.8 |
| Average | 1.7 | 1.8 |

After (Kesteren et al. 2000)

Sadly, it seems that my new adopted home (Australia) has one of the highest burglary rates according to this data and has maintained this for some time. Looking at the data from the 1988 and 1991 a study suggested that Australia had one of the highest victimisation rates for burglary in both of these years with a combined-year victimisation rate of 4.0% for burglary and 3.8% (Van Dijk and Mayhew 1992). There was very little change in the 2000 data (Table 9.2) although the attempted burglary rate appeared to reduce reasonably significantly whereas the actual burglary rate increased by a small amount (0.1%). What is quite telling about this data is that in the intervening period (relatively) strict gun legislation was introduced by John Howard, then Prime Minister of Australia[5]. This was in response to the 29 April 1996 Port Arthur mass shootings that resulted in the deaths of 35 people and the injury of another 23 people. Howard introduced the "National Firearms Buyback Program", which ran for 12 months and retrieved around 650,000 guns. Another shooting at a University campus on 21 October 2002 prompted a further buy-back program. The 2003 handgun buyback ran for 6 months and retrieved 69,000 handguns. If indeed more guns would result in less burglaries, then Australia would have arguably seen a

---

[5]It is notable that Howard was a newly elected conservative Prime Minister and brought about a National Firearms Agreement that encompassed all States and Territories in Australia in twelve days after the massacre. This is now held up as a model for gun control across the world.

substantial increase in attempted and actual burglary rates. This simply was not the case. In fact, by 2003/2004 this data had dropped further.

Gun ownership in the home has also been found to correlate with an increased risk of murder by a family member or close acquaintance (Kellermann et al. 1993). The presence of a gun in the home is also risk factor for femicide (where a male family member murders a female family member on account of her gender). In fact, the evidence for this is overwhelming (Campbell et al. 2003; Campbell et al. 2007; Bailey et al. 1997; Wiebe 2003). So, statistically speaking it is not sensible to keep a gun in the home.

### 9.3.3  To Stop a Bad Guy with a Gun It Takes a Good Guy with a Gun

There is no doubt that in some circumstances this is true and was the globally broadcast mantra of the NRA head, Wayne LaPierre in the aftermath of the Florida School shooting on 14 February 2018 (BBC 2018). Nineteen-year-old Nikolas Cruz murdered seventeen students and staff with an AR-15 assault rifle before dropping the rifle and his bullet-proof vest to blend in with other students. After this shooting there was much discussion on the arming of schoolteachers, a view that President Trump supported. And there is no doubt that a good guy with a gun will stop a bad guy with a gun. After all, that is why we arm our police forces. So perhaps 'more guns' is the answer. However ironically in this case it was not! An armed deputy police officer arrived at the scene during the shootings and failed to proceed to the direction of Cruz citing that he believed that the shooting was been carried out outside and the school was operating lock-down procedures.

Whatever the truth in this matter there are a number of potential problems arming teachers. Firstly, we expect teacher to teach and not to shoot. Teachers pursue that profession for the love of conveying new information to eager young minds. Secondly, it is questionable whether teachers would be sufficiently responsive in the event of a perpetrator with gun forcing themselves into the classroom. Even seasoned police officers and army veterans who have substantial experience with firearms can find these circumstances difficult!

### 9.3.4  Gun Ownership in the Home Correlates with Gun-Induced Suicide

There is a general belief that limiting access to firearms could prevent many suicides. However, this is often controversial with opponents citing opposing scientific studies on the subject. Nevertheless, the evidence appears to point to a fact that easy access to firearms increases the risk of death by suicide. However, we must be somewhat cautious here. The important question is not whether gun access increases the risk of gun suicide (that conclusion is pretty obvious!) but whether easy access to a gun will increase the *total number* of suicide victims in society.

The statistical evidence appears to point to the latter. A population cohort comprising of 238,292 who purchased a handgun in California in 1991 were studied for a period of approximately 6 years from 15 days after the gun was purchased through 31 December 1996. The researchers concluded that the purchase of a handgun is associated with a substantial increase in the risk of suicide by firearm and by any method (meaning that the overall rate of suicide increased with gun owner-ship). The increase in the risk of suicide by firearm is apparent within a week after the purchase of a handgun and persists for at least six years (Wintemute et al. 1999). In fact, there was a staggering increase in the suicide rates soon after the purchase of a gun implying that in some instances, guns were purchased for the sole purpose of committing suicide. Arguably, forced waiting periods or "cooling off" periods would lead to a reduction in overall suicides.

Further, in a 1992 study, Kellermann and co-researchers published a study in the New England Journal of Medicine (Kellermann et al. 1992) to evaluate the strength of correlation between gun ownership and suicide. Over a 32-month period 803 suicides were reported in two counties: Shelby County, Tennessee, and King County, Washington. Of the suicides that occurred in the home (565) roughly 58% were committed with a firearm. After a statistical comparison to a control group, they concluded that the presence of one or more guns in the home was found to be associated with an increased risk of suicide.

There have been many studies that point to similar conclusions e.g., (Cummings et al. 1997; Miller et al. 2002). A further study was published in 2016 that noted a strong correlation between gun ownership and firearm-induced suicide (Siegel and Rothman 2016). The researchers noted that in males, easier access to guns would result in more overall suicides whereas the same pattern was not seen in females. Their conclusion suggested that policies that reduce firearm ownership will likely reduce suicides by all means and by firearms. For female persons, such policies will likely reduce suicides by firearms (but not overall suicide rates).

Probably the most in-depth survey of this issue was published by the National Academy of Sciences in *Firearms and Violence: A Critical Review*. This was based on a committee's review of all of the available published research data. Their conclusions were pretty clear in that "States, regions, and countries with higher rates of household gun ownership have higher rates of gun suicide" . Further they noted that there was a "more modest" association between gun ownership and the overall risk of suicide. They noted that the reason for this was unclear (Wellford et al. 2005).

This also correlates with suicide data from other wealthy nations (although again we must be somewhat cautious here as there are many factors that can affect suicide rates that exclude the method). The US currently has one of the highest suicide rates when compared to other OECD countries with recent data showing suicide rates of 13.9 per 100,000 of the population compared to the OECD average of 11.5 (Tikkanen and Abrams 2020).

Of course, mental anguish and people with suicidal tendencies can act impulsively and rashly to end their life and this is perhaps the most convincing argument against making guns so easily accessible in the home. It does not take much force to pull a trigger when a loaded gun is pointing at your head. There is not much effort required. And, the end would come in a mostly instantaneous fashion. I say 'mostly', as there have been a staggering number of suicide attempts that have resulted in brain damage or permanent disablement. The sad fact is that with a gun there is no second chance.

The really important question to answer here however, is do gun control measure make a difference? Strikingly it was noted in the report *Firearms and Violence: A Critical Review*: "Some gun control policies may reduce the number of gun suicides, but they have not yet been shown to reduce the overall risk of suicide in any population". This seems counter-intuitive. However some control measures may result in unintended consequences. For example, mental health advocates have opposed registries of persons with a history of mental illness, arguing that the stigma of appearing in a Government-sponsored registry could lead some persons to refuse the much-needed mental health treatment. This could potentially increase, rather than decrease the risk of a lethal outcome (Wellford et al. 2005).

### 9.3.5   Restricting Access to Guns Would Lead to Less 'Spur of the Moment' Crimes

On 12 October 2018 Helen Washington, 75, shot her grandson for ignoring her repeated demands to keep his teacup off her furniture. Police arrived to Washington's Brooklyn Center, Minnesota, home to find her unnamed grandson in the front yard with a gunshot wound to his thigh, claiming his grandmother was the shooter. Would gun control have stopped this bizarre crime? Possibly. There are many examples like this that crop up.

Time to time we are confronted with example where, on the surface, sane, normal intelligent people do the bizarre. And this is impossible to predict. In fact, the evidence points to this. A study what looked at 400 homicides in the United States showed that 44% of those homicides arose as a result of an altercation or quarrel (Kellermann et al. 1993). There is a harsh finality with guns, that in a split-second life can be extinguished. There was a time when disputes were settled with a fistfight, and yes, each participant was hurt and badly bruised but they would have been able to walk away from the altercation. In fact, from my personal teenage experience I found it resulted in new friends being made. That is, my nemesis became my best friend once we had had a fight and subsequently made up.

The harsh finality of firearms that is delivered in such a forceful and rapid manner is probably best illustrated by the speed of the projectile that leaves the muzzle. At a range of 0.5 m, a typical hand-gun bullet will take around 1.5 ms to reach its target. That is a tiny fraction of the time it takes for the eye to blink (which takes around 300–400 ms). Even at a range of around 50 m a blinking eyelid will be just about shut as the projectile slams into the target. So, of course, to dodge a speeding bullet is impossible.

## 9.4   So, What Is the Solution?

There is no doubt, that there is a strong correlation between gun ownership and the homicide-rate in wealthy countries with the US leading the way in that regard. Therefore, where the gun control has been legislated and accepted by the society (the UK, for example) it only seems reasonable to maintain that level of control. Nevertheless, where there is strong, and somewhat polarised, public support for guns (such as in the US) there has to be another way.

I like the view that has been promoted by Hemmingway and others, which essentially suggests that we should treat gun-inflicted injury as a public health problem much in the same way we society has done with smoking-inflicted disease, automobile accidents, air pollution and so on. The key is clearly to place the burden on manufacturers to come with new designs that can reduce mortality. In a day where we appear to be inundated with risk assessments and with the aim of making everything safer there has clearly been little progress in making guns safer. Technology, perhaps, may hold the answer.

Take the example of motor vehicles. In the UK, from 1942 to 1965 the number of casualties from road-traffic accidents steadily increased along with the number of vehicles on the roads. In 1966 the number of fatalities on Britain's roads peaked at 7,985. Since then there has been a steady decrease in fatalities despite the number of vehicles on the roads increasing dramatically (there were approximately 4 million vehicles on Great British roads in 1950 which steadily grew to over 34 million in 2010). So why the decrease in fatalities? There was a substantial push to reduce traffic deaths from 1966 with the introduction of drink-driving laws. Other factors that have played their part to reduce fatalities are legal requirements for wearing seatbelts, speed cameras and improved road design. It is also not too difficult to see that improvements in vehicle designs have also helped and savvy drivers are now more than ever seeking cars that have the highest EURONCAP rating. Airbags, active breaking systems, stability assist, anti roll cages, energy absorbing crush tubes, crumple zones, are pretty much standard for all modern automobiles. All of these factors help to improve the survivability of the occupant.

Other products go through similar processes and there is an industrial imperative to increase the safety of products. Manufacturers strive to make lawnmowers safer by virtue of automatic cut-outs to prevent runaway machines. Nail guns have evolved so that it is virtually impossible to injure somebody intentionally. And yet, it seems that gun manufacturers are immune for this increased push for safety. Why is this? Why is it for example that it is usually impossible to tell if a round is chambered (a frequent reason for accidental injury) in a gun and yet with my son's Nerf gun, you can? Or take my daughter's polaroid camera. How is it she can see that there is a film present before opening the film compartment? Yet, this seems to allude most gun makers.

### 9.4.1  Pre-purchase Tests

Some have suggested that one of the solutions for reducing gun violence in society may be in strict pre-purchase tests. In certain countries gun ownership is rare, simply because of the pre-purchase tests, and some advocates have argued that this is the solution. For example, in Japan which maintains very strict gun laws, these are the steps:

- First you take a firearms class and written exam
- Then you go to your doctor for a mental health exam
- Then you must apply for shooting lessons
- You then need to explain to police why you need a gun
- If the police are satisfied, then you must pass a personal criminal history review
- Then you must obtain a gunpowder permit
- You must take a one-day training test including passing a firing test
- Then you need a certificate for the gun you want
- You will also be required to buy a safe for your gun that meets strict safety regulations
- Then the police must inspect your gun case and follow with another background review.

Japan has one of the lowest gun-crime rates in the world with generally fewer than 10 gun-deaths per year in a population that exceeds 125 million.

### 9.4.2  Gun Buy-Back Schemes and Amnesties

Gun buy-back schemes and amnesties are designed to reduce the level of gun owner-ship in society. This was carried out in Australia after the 1996 Port Arthur massacre. On April 28, 1996 the Howard government brought in a massive buy-back scheme and banned semi-automatic and automatic rifles. We have already looked at how Australia's buy-back schemes in 1996 and 2003 affected burglary rates, but what about gun-related deaths?

There have been several studies that looked at the effect of the gun buy-back schemes. Different views are held but there is a large swathe of evidence to suggest that the gun buy-back programs in Australia worked. It is claimed in the literature that the gun buyback program significantly reduced Australia's homicide rate (Bartos et al. 2020; Leigh and Neill 2010); suicide rates (Ozanne-Smith et al. 2004) and that it resulted in less mass shootings (Chapman et al. 2006). There had been multiple mass shootings up to 1997 but no mass shootings from 1997 to 2016 (Chapman et al.

2016; Lemieux et al. 2015). Notably it is often cited that there were 13 fatal mass shootings occurred in Australia from 1979–1996. However, post gun law reforms there have been none. True, this ignores the Monash shootings in 2003 and the Lindt Café siege in Sydney in 2014. However, this is due to the fact that the researchers used a definition of mass shooting that was more than, or equal to five victims killed by gunshot, not including the perpetrator. The aforementioned atrocities resulted in fewer deaths than five. Based on this definition, it has been claimed that:

> "Before 1996, approximately 3 mass shootings took place every 4 years. Had they continued at this rate, approximately 16 incidents would have been expected since then by February 2018" (Chapman et al. 2018).

Thankfully, this has not been the case. It has also been shown that the States and Territories that bought back more guns per 100,000 of the population showed a greater reduction in firearm suicides and homicides (Leigh and Neill 2010). Notably, there was also an increase in non-firearm suicides for the areas that bought back more guns, implying that the victim simply migrated to other methods with the absence of a gun.

Other seem to disagree in that there was a limited impact or even no impact at all (Baker and McPhedran 2007; Lee and Suardi 2010). These conclusions have been largely based upon the fact that the firearm homicide and suicide rates had been dropping in the 1990s anyway. However, McPhedran and Baker have not been without their critics. In particular their 2007 paper published in *The British Journal of Criminology* has been criticised for using a reduced time-series analysis (Neill and Leigh 2007; Hemenway 2011). Further in a paper published in 2011, McPhedran and Baker claimed that:

> It does not find support for the hypothesis that Australia's prohibition of certain types of firearms has prevented mass shootings, with New Zealand not experiencing a mass shooting since 1997 despite the availability in that country of firearms banned in Australia. (McPhedran and Baker 2011).

Firstly, given the rarity of such horrific events overall, it is difficult to compare New Zealand to Australia given that New Zealand has approximately the population of Sydney. Secondly, New Zealand has recently had a deadly mass shooting. On 15 March 2019, 51 worshippers were killed in the Christchurch mosque shootings prompting the Ardern Government to introduce restrictive gun laws.

Some of these aforementioned authors from both sides of the argument, it seems have competing affiliations (Frame 2019). Nevertheless, in Australia it did appear that the gun buyback scheme worked in reducing the number of firearms-related deaths.

### 9.4.3   Smart Guns

Judge Dredd was one of my favourite comic-book heroes growing up. Dredd had a notable firearm in that only he could fire his weapon and woe-betide anybody else that tried to fire *his* weapon.

Today, we are inundated with smart phones, smart cars, and even smart houses. So, why not smart guns?

Smart guns are guns that can be unlocked by virtue of a fingerprint (as is commonly used on smart phones) or by the use of a radio frequency transmitting device that is stored in a ring or bracelet that is matched to the gun. I suppose one of the fascinating aspects of gun ownership is that, by and large, guns are one piece of technology that has escaped technological improvements over the past 100 years. Maybe it is time to change that!

### 9.4.4   The Last Word

Guns can be exciting, but we all know they are dangerous. The debate continues to rage on whether or not there are benefits in gun control. Most people would see that it would be nonsense to provide everybody in society access to whatever weapon they would wish for. We simply do not. We do not allow access to biological weapons, anti-tank guided systems or tactical nuclear munitions. Why would we? So, there must be *some* sensible restrictions in place. Disagreements on how restrictive gun control should be have their merits and ultimately this debate can only be resolved politically. This issue is about the power of the state and the rights of the individual. The state principally has the responsibility to protect wider society whereas the individual has the responsibility to behave as good citizens and to protect themselves and their loved ones. To quote Donald Rumsfeld, there are two 'known unknowns' here. Does the state have the authority to enforce gun control? And, does controlling access make any appreciable difference? The answer to the first question in many countries is 'yes'. For the second question, the problem is that the science is not crystal clear for a variety of reasons and the answer is different from country to country. Each country is unique because its culture and political system is different. What is acceptable and what works in one country may not work in another. So, possibly, the debate will continue to rage.

## References

Appelbaum PS (2013) Public safety mental disorders and guns. JAMA Psychiatry 70(6):565–566. https://doi.org/10.1001/jamapsychiatry.2013.315

Bailey JE, Kellermann AL, Somes GW, Banton JG, Rivara FP, Rushforth NP (1997) Risk factors for violent death of women in the home. Archiv Internal Med 157(7):777–782. https://doi.org/10.1001/archinte.157.7.777

Baker J, McPhedran S (2007) Gun laws and sudden death: did the Australian firearms legislation of 1996 make a difference? British J Criminol 47(3):455–469. https://doi.org/10.1093/bjc/azl084

Bartos BJ, McCleary R, Mazerolle L, Luengen K (2020) Controlling gun violence: assessing the impact of Australia's gun buyback program using a synthetic control group experiment. Preven Sci 21(1):131–136. https://doi.org/10.1007/s11121-019-01064-8

BBC (2018) Seven things the NRA blames after Florida. NRA head: gun control advocates 'exploiting' Florida tragedy. www.bbc.com

Camerer CF, Dreber A, Holzmeister F, Ho T-H, Huber J, Johannesson M, Kirchler M, Nave G, Nosek BA, Pfeiffer T, Altmejd A, Buttrick N, Chan T, Chen Y, Forsell E, Gampa A, Heikensten E, Hummer L, Imai T, Isaksson S, Manfredi D, Rose J, Wagenmakers E-J, Wu H (2018) Evaluating the replicability of social science experiments in nature and science between 2010 and 2015. Nature Human Behav 2(9):637–644. https://doi.org/10.1038/s41562-018-0399-z

Campbell JC, Glass N, Sharps PW, Laughon K, Bloom T (2007) Intimate partner homicide: review and implications of research and policy. Trauma Violence Abuse 8(3):246–269. https://doi.org/10.1177/1524838007303505

Campbell JC, Webster D, Koziol-McLain J, Block C, Campbell D, Curry MA, Gary F, Glass N, McFarlane J, Sachs C, Sharps P, Ulrich Y, Wilt SA, Manganello J, Xu X, Schollenberger J, Frye V, Laughon K (2003) Risk factors for femicide in abusive relationships: results from a multisite case control study. Amer J Public Health 93(7):1089–1097. https://doi.org/10.2105/AJPH.93.7.1089

Chapman S, Alpers P, Agho K, Jones M (2006) Australia's 1996 gun law reforms: faster falls in firearm deaths, firearm suicides, and a decade without mass shootings. Injury Preven 12(6):365–372. https://doi.org/10.1136/ip.2006.013714

Chapman S, Alpers P, Jones M (2016) Association between gun law reforms and intentional firearm deaths in Australia, 1979–2013. JAMA-J Amer Med Assoc 316(3):291–299. https://doi.org/10.1001/jama.2016.8752

Chapman S, Stewart M, Alpers P, Jones M (2018) Fatal firearm incidents before and after Australia's 1996 national firearms agreement banning semiautomatic rifles. Annals Internal Med 169(1):62–64. https://doi.org/10.7326/m18-0503%m29532070

Cummings P, Koepsell TD, Grossman DC, Savarino J, Thompson RS (1997) The association between the purchase of a handgun and homicide or suicide. Amer J Public Health 87(6):974–978. https://doi.org/10.2105/AJPH.87.6.974

Douglas J, Burgess AW, Burgess AG, Ressler RK (2006) In: Crime classification manual: a standard system for investigating and classifying violent crimes. Wiley

Fox JA, DeLateur MJ (2014) Mass shootings in America: moving beyond Newtown. Homicide Stud 18(1):125–145. https://doi.org/10.1177/1088767913510297

Frame T (2019) Gun control: what Australia got right (and wrong). UNSW Press, Sydney, Australia

Hemenway D (2011) The Australian gun buyback, vol 4. Joyce Foundation, Harvard, USA

Hemenway D (2017) Private guns public health. University of Michigan Press, Ann Arbor, USA

Karp A (2018) Estimating global civilian-held firearms numbers—briefing paper. Small Arms Survey, Geneva, Switzerland

Kellermann AL, Rivara FP, Rushforth NB, Banton JG, Reay DT, Francisco JT, Locci AB, Prodzinski J, Hackman BB, Somes G (1993) Gun ownership as a risk factor for homicide in the home. New England J Med 329(15):1084–1091. https://doi.org/10.1056/NEJM199310073291506

Kellermann AL, Rivara FP, Somes G, Reay DT, Francisco J, Banton JG, Prodzinski J, Fligner C, Hackman BB (1992) Suicide in the home in relation to gun ownership. New England J Med 327(7):467–472

Kesteren Jv, Mayhew P, Nieuwbeerta P (2000) Criminal victimisation in seventeen industrialised countries: key findings from the 2000 International Crime Victims Survey. Justitie Wetenschappelijk Onderzoek- en Documentatiecentrum, The Hague

Kopel DB (2013) The truth about gun control. In: Encounter broadsides, Encounter Books, vol 32. New York, USA

Krouse WJ, Richardson DJ (2015) Mass murder with firearms: incidents and victims 1999–2013. CRS Report. Congressional Research Service, Washington DC

Lankford A (2016) Public mass shooters and firearms: a cross-national study of 171 countries. Violence Vict 31(2):187–199. https://doi.org/10.1891/0886-6708.Vv-d-15-00093

Lee WS, Suardi S (2010) The Australian firearms buyback and its effect on gun deaths. Contemporary Econ Policy 28(1):65–79. https://doi.org/10.1111/j.1465-7287.2009.00165.x

Leigh A, Neill C (2010) Do gun buybacks save lives? Evidence from panel data. Amer Law Econ Rev 12(2):509–557. https://doi.org/10.1093/aler/ahq013

Lemieux F, Bricknell S, Prenzler T (2015) Mass shootings in Australia and the United States, 1981-2013. J Criminolog Res Policy Practice 1(3):131–142. https://doi.org/10.1108/JCRPP-05-2015-0013

Lott JR (2018) How a botched study fooled the world about the U.S. share of mass public shootings: U.S. Rate is lower than global average. Social Science Research Network (28 September 2018). http://dx.doi.org/10.2139/ssrn.3238736

McPhedran S, Baker J (2011) Mass shootings in Australia and New Zealand: a descriptive study of incidence (2008). Justice Policy J 8

Miller M, Azrael D, Hemenway D (2002) Firearm availability and unintentional firearm deaths, suicide, and homicide among 5–14 year olds. J Trauma 52(2):267–275. https://doi.org/10.1097/00005373-200202000-00011

Nand D, Naghavi M, Marczak LB, Kutz M, Shackelford KA, Arora M, Miller-Petrie M, Aichour MTE, Akseer N, Al-Raddadi RM, Alam K, Alghnam SA, Antonio CAT, Aremu O, Arora A, Asadi-Lari M, Assadi R, Atey TM, Avila-Burgos L, Awasthi A, Quintanilla BPA, Barker-Collo SL, Bärnighausen TW, Bazargan-Hejazi S, Behzadifar M, Bennett JR, Bhalla A, Bhutta ZA, Bilal AI, Borges G, Borschmann R, Brazinova A, Rincon JCC, Carvalho F, Castañeda-Orjuela CA, Dandona L, Dandona R, Dargan PI, De Leo D, Dharmaratne SD, Ding EL, Do HP, Doku DT, Doyle KE, Driscoll TR, Edessa D, El-Khatib Z, Endries AY, Esteghamati A, Faro A, Farzadfar F, Feigin VL, Fischer F, Foreman KJ, Franklin RC, Fullman N, Futran ND, Gebrehiwot TT, Gutiérrez RA, Hafezi-Nejad N, Bidgoli HH, Hailu GB, Haro JM, Hassen HY, Hawley C, Hendrie D, Híjar M, Hu G, Ilesanmi OS, Jakovljevic M, James SL, Jayaraman S, Jonas JB, Kahsay A, Kasaeian A, Keiyoro PN, Khader Y, Khalil IA, Khang YH, Khubchandani J, Kiadaliri AA, Kieling C, Kim YJ, Kosen S, Krohn KJ, Kumar GA, Lami FH, Lansingh VC, Larson HJ, Linn S, Lunevicius R, Abd El Razek HM, Abd El Razek MM, Malekzadeh R, Malta DC, Mason-Jones AJ, Matzopoulos R, Memiah PTN, Mendoza W, Meretoja TJ, Mezgebe HB, Miller TR, Mohammed S, Moradi-Lakeh M, Mori R, Nand D, Nguyen CT, Le Nguyen Q, Ningrum DNA, Ogbo FA, Olagunju AT, Patton GC, Phillips MR, Polinder S, Pourmalek F, Qorbani M, Rahimi-Movaghar A, Rahimi-Movaghar V, Rahman M, Rai RK, Ranabhat CL, Rawaf DL, Rawaf S, Rowhani-Rahbar A, Safdarian M, Safiri S, Sagar R, Salama JS, Sanabria J, Milicevic MMS, Sarmiento-Suárez R, Sartorius B, Satpathy M, Schwebel DC, Seedat S, Sepanlou SG, Shaikh MA, Sharew NT, Shiue I, Singh JA, Sisay M, Skirbekk V, Filho AMS, Stein DJ, Stokes MA, Sufiyan MB, Swaroop M, Sykes BL, Tabarés-Seisdedos R, Tadese F, Tran BX, Tran TT, Ukwaja KN, Vasankari TJ, Vlassov V, Werdecker A, Ye P, Yip P, Yonemoto N, Younis MZ, Zaidi Z, El Sayed Zaki M, Hay SI, Lim SS, Lopez AD, Mokdad AH, Vos T, Murray CJL (2018) Global mortality from firearms, 1990–2016. JAMA-J Amer Med Assoc 320(8):792–814. https://doi.org/10.1001/jama.2018.10060

Neill C, Leigh A (2007) Weak tests and strong conclusions: a re-analysis of gun deaths and the Australian firearms buyback (June 2007). ANU, Centre for Economic Policy Research, Canberra, Australia

Ozanne-Smith J, Ashby K, Newstead S, Stathakis VZ, Clapperton A (2004) Firearm related deaths: the impact of regulatory reform. Injury Preven 10(5):280–286. https://doi.org/10.1136/ip.2003.004150

Saad L (2013) Americans Fault Mental Health System Most for Gun Violence. Princeton, NJ

Siegel M, Rothman EM (2016) Firearm ownership and suicide rates among US Men and Women, 1981–2013. Amer J Public Health 106(7):1316–1322

Swanson JW, McGinty EE, Fazel S, Mays VM (2015) Mental illness and reduction of gun violence and suicide: bringing epidemiologic research to policy. Annals of Epidemiol 25(5):366–376. https://doi.org/10.1016/j.annepidem.2014.03.004

Tikkanen R, Abrams MK (2020) U.S. Health Care from a Global Perspective, 2019: Higher Spending, Worse Outcomes? Commonwealth Fund, New York, USA

Van Dijk JJM, Mayhew P (1992) Criminal victimization in the Industrialized World: Key findings of the 1989 and 1992 International Crime Surveys. Ministry of Justice, Department of Crime Prevention, The Hague, The Hague

Wellford CF, Pepper JV, Petrie CV (2005) Firearms and violence: a critical review

Wiebe DJ (2003) Homicide and suicide risks associated with firearms in the home: a national case-control study. Annals of Emerg Med 41(6):771–782. https://doi.org/10.1067/mem.2003.187

Wintemute GJ, Parham CA, Beaumont JJ, Wright M, Drake C (1999) Mortality among recent purchasers of handguns. New England J Med 341(21):1583–1589. https://doi.org/10.1056/NEJM199911183412106

# Glossary

**Adiabatic shear** A process where, during the penetration of a projectile in a target, the rate of thermal softening exceeds the rate of work hardening leading to the formation of shear failure bands. There is no heat transfer from (or to) the heat-affected area

**Alloy** A mixture of two or more elements where at least one of them is a metal

**Ammonia** a colourless gas with a characteristic pungent smell, which dissolves in water to give a strongly alkaline solution

**Ammunition** Any munitions of war whether filled with solid shot or explosive

**Amorphous** A non-crystalline state where the arrangement of atoms has no regular order

**Angle of obliquity** The angle between the projectile trajectory and the normal to the surface of impact

**Anti-tank guided weapon** A vehicle or infantry launched warhead that is capable of being guided or guiding itself to attack a target—usually with a shaped-charge warhead

**Armoured fighting vehicle** The generic name for a military vehicle, tracked or wheeled, that is designed to engage in warfighting

**Armoured personnel carrier** An armoured vehicle (tracked or wheeled) designed to carry troops into a conflict zone

**Armour piercing** Projectile with a hard core designed to penetrate and perforate hard targets

**Armour piercing discarding sabot** A sub-calibre solid shot of relatively low length to diameter ratio (~5) that is carried up a gun barrel by a sabot that is discarded when exiting the muzzle. It is usually spin stabilised

**Armour piercing fin stabilised discarding sabot** A sub-calibre solid shot of relatively high length to diameter ratio (~15) that is carried up a gun barrel by a sabot that is discarded when exiting the muzzle. It is usually drag stabilised

**Armour piercing incendiary** Projectile with an armour-piercing core and a low-explosive material encased in the tip that deflagrates when the bullet impacts armour

© Springer Nature Switzerland AG 2021
P. J. Hazell, *The Story of the Gun*, Springer Praxis Books,
https://doi.org/10.1007/978-3-030-73652-1

**Arquebus** An early form of medieval gun, that would have been supported on a tri-pod

**Attenuate** To reduce in force or value. The term is frequently used to describe the weakening of a stress or shock wave

**Austenite** A soft crystalline structure that is formed in steels at elevated temperatures. It is occasionally present with martensite in drastically quenched steels

**Automatic action rifle** A fully automatic rifle that loads fires and ejects the cartridge case as along as the trigger is depressed

**Azimuth** A directional bearing but usually used in reference to protection in terms of a 90° arc—that is, from the zenith to the horizon

**Bainite** A constituent of steel that can form when a steel is cooled from the austenitic temperature. Under a microscope it can be seen as a fine plate-like structure of cementite and ferrite that is very strong. It is named after Edgar Collins Bain (1891–1971)

**Ballistic limit velocity** The velocity at which there is a specific probability that a known projectile will just perforate the target. Therefore a $v_{50}$ is the velocity at which 50% of the projectiles will just perforate the target

**Ballistic pendulum** A device for measuring the momentum of a projectile or fragment(s)

**Behind armour effects** The fragmentation, or blast and over pressure that occurs after a projectile or shaped-charge jet perforates armour

**Blast wave** A destructive wave produced by the detonation of an explosive

**Blunt trauma** Injury that occurs when a body armour vest is not perforated but the momentum transfer of the impact causes large deformation in the backing layer. It can lead to bruising, serious injury to major organs or even death

**Bolt-action-rifle** A rifle that uses a steel cylindrical 'bolt' that locks behind the cartridge case to prevents its rear movement and is manually unlocked usually by a rotating movement

**Brass** A metal alloy that is yellow in colour that comprises of copper and zinc

**Brinell indentor** A spherical hard steel indentor that is used to measure the hardness of a material—usually metal

**Brittle fracture** This occurs when a projectile strikes a target with a low *fracture toughness*. Typically, the target will shatter into a large number of fragments. Examples of materials that suffer brittlefrom fracture are brick, ceramic, and glass. Very little energy is required to form new fracture surfaces

**Bronze** A metal alloy that is yellowish-brown in colour that comprises of copper and tin

**Built-up gun** A gun that is constructed from tubes upon which outer jackets or hoops are shrunk fit—sometimes with wire winding inbetween

**Bullet** A projectile that is fired from a gun and is usually encased in a metallic jacket that engages with the gun's rifling to enable spin stabilisation

**Bump stock** A device that is used to replace the existing stock of a semi-automatic weapon and literally 'bumps' the whole gun forward such that the trigger is pushed forward onto the finger of the shooter

**Calibre** The nominal diameter of the bore of a gun. For a rifled barrel it is measured across the lands of the rifling

**Cannelure** A Cannelure groove around a bullet into which the edge of the cartridge case is crimped

**Carbine** A shortened rifle

**Carburising** The diffusion of carbon into a material from a carbon-rich environment with the application of heat

**Cartridge case** A container that holds the primer and propellant and interfaces with the projectile, usually made from brass

**Cavitation** Formation of an empty space. Within the context of wound ballistics, it is described in terms of the formation of temporary cavity within the human body

**Cementite** A constituent in steel that is very hard and brittle. Also known as iron carbide, it is a compound of iron and carbon and has the chemical formula of $Fe_3C$

**Ceramic** A solid compound that is formed by the application of heat, and sometimes heat and pressure, comprising at least one metal and non-metallic elemental solid

**Cermet** A composite material composed of ceramic and metallic materials. A metal is used as a binder for an oxide, boride or carbide (for example, tungsten carbide). The metallic elements normally used are nickel, molybdenum, and cobalt

**Charge** Enclosed quantity of high explosive or propellant with its own integral means of ignition

**Chase of the gun** The sloping portion forward of the breech extending to the muzzle. Usually used in the context of 'built-up' guns

**CJ point** With reference to a detonating high explosive material, this is the pressure of the detonation products at the cusp of the detonation wave. Named in honour of David Chapman (1869–1958) and Jacques Charles Émile Jouguet (1871–1943)

**Coefficient of thermal expansion** The change in size of a material with the change in temperature. It has the units of 1/K or 1/°C

**Columbiad** A large-calibre, smoothbore, muzzle-loading cannon used to fire heavy projectiles

**Combat body armour** A UK body armour that consists of an aramid and nylon construction to which ceramic-faced armour can be applied as an insert to protect the major organs

**Complete penetration** See *perforation*

**Composite** A structure comprising of two or more materials often engineered so that the properties of the materials are complementary and therefore the structure is more than the sum of its parts

**Coriolis** An inertial force that results in the deflection of a projectile due to the spin of the Earth

**Crystalline** A material that possesses a structure that consists of an ordered array of atoms

**Delamination** With reference to a composite armour system it is the process by which the individual layers (or laminae) become separated from one another—usually due to the penetration or perforation of a projectile

**Detonation wave** A shock front supported by the presence of chemical energy

**Disruptor** The part of the armour system that causes the projectile to fracture and fragment. Usually these are made from materials of high hardness and/or impedance

**Driving band** A malleable or pre-engraved band that is pressed around the rim of the projectile which, when engaged with the rifling of the barrel, imparts spin to the projectile

**Ductile fracture** The growth and coalescence of voids within a ductile medium under stress such that separation of the material is inevitable

**Ductile hole growth** With reference to a projectile penetrating a ductile target, this occurs when the material plastically deforms such that the material is pushed out of the way of the penetrator. It is important to realise that no localised increase in density occurs around the hole formed from the penetrator rather the whole plate deforms to take into account the hole that is formed—that is, the volume remains constant

**Elastic** A material is said to be elastic if it returns back to its original shape after being stretched or squeezed

**Equation of state** A fundamental constitutive equation that relates a material's response to pressure in terms of density and internal energy

**Exothermic** A reaction accompanied by the release of heat

**Explosively-formed projectile** A projectile that is formed from a thin geometric shape (usually a 'dish') by the action of a detonating high explosive. Also sometimes referred to as a 'self-forging fragment'

**Explosive reactive armour** An armour system that works be intercepting a penetrator or shaped-charge jet with explosively driven plates in order to cause disruption

**Face hardening** The process of hardening the surface of a material—usually by work hardening, flame hardening, or carburising

**Flame hardening** A process of heating the surface of steel (by using a gas flame) up to very high temperatures and then rapidly cooling (quenching) to form a very hard but brittle layer with decreasing hardness through the thickness of the steel plate

**Flexural rigidity** The multiplication of the Young's modulus and second moment of area

**Flint** A hard mineral, usually dark grey, that was used in flintlock mechanisms to generate sparks through mechanical impact with steel

**Flintlock** A type of gun, or mechanism on a gun, where the ignition of the propellant occurred by the sparks generated by the mechanical impact of a flint by a steel hammer

**Fracture** The separation of material

**Fracture toughness** A term that defines a material's ability to resist the extension of a crack when the material is placed under load

**Glacis plate** The sloping plate of armour located at the front and top of an AFV

**Grain** A single crystal of material. Polycrystalline materials exhibit multiple grains

**Gross cracking** If on impact by a projectile a crack grows in a plate, that crack will propagate in a manner similar to brittle fracture (depending on the material's toughness). The crack will propagate at a high velocity and cause a large portion of the material to become separated from the bulk. High-hard armour and some hardened aluminium alloys can be susceptible to gross cracking when penetrated

**Hardness** A measure of resistance to indentation, abrasion and wear

**Headspace** The distance measured from the face of the bolt to the part of the chamber that stops forward motion of the cartridge. Applicable in small arms terminology

**High-entropy alloys** Alloys that are formed by mixing roughly equal of five or more elements

**Hydrodynamic penetration** Penetration of a projectile into a target material where the penetrator and target behave as if they possess no strength. This tends to occur at elevated velocities of impact where the shock pressures generated are orders of magnitude higher than the strengths of materials

**ICBM** Intercontinental Ballistic Missile—a missile capable of traversing large distances, usually with a nuclear payload

**Impulse** The change of momentum of an object. For an object in motion, it can also be calculated by the applied force multiplied by the time over which that force is acting on the object

**Inelastic** Not elastic—that is, plastic

**Internal energy** Energy that a material possesses because of the motion its atoms and molecules

**Interface defeat** Resistance to penetration of a hard target due to the penetrator being completely eroded at the target's surface via dwell

**Iron Carbide** See cementite

**Isotropic** Possessing identical material properties in all directions in three-dimensional space

**Kinetic energy** Energy due to mass and velocity. It can be calculated by taking half the mass and multiplying it by the velocity squared ($\frac{1}{2}\,m\,v^2$)

**Long-rod penetrator** A solid rod-like shot, usually sub-calibre and drag stabilised, which is used in attacking armour (see *Armour Piercing Fin Stabilised Discarding Sabot*)

**Magnus effect** The effect that describes the deflection of a spinning projectile transiting through air—such as a spinning soccer or ping-pong ball named after Heinrich Gustav Magnus (1802–1870)

**Matchlock** A mechanism whereby a lever carrying a smouldering wick was lowered into a flash-pan carrying gunpowder at the pull of the trigger

**Martensite** A fine needle-like structure that is the hardest constituent obtained in steel

**Microcrack** A very small crack that may or may not propagate depending on the stress concentration at its tip

**NRA** National Rifle Association

**OECD** Organization for Economic Cooperation and Development—an intergovernmental economic organisation with 37 member countries founded in 1961 to stimulate economic progress and world trade

**Passive armour**  Armour that works to defeat an incoming projectile or shaped-charge jet by mechanical properties

**Partial penetration**  Not *perforation*

**Penetration**  The process of a projectile moving through a material

**Penetrator**  A projectile that penetrates

**Penetrator dwell**  The process of a projectile being unable to penetrate the ceramic until its strength has been diminished. The penetrator therefore appears to dwell on the surface of the target. If the strength of the ceramic does not diminish in the timescale of penetrator erosion then interface defeat occurs

**Perforation**  The process of the projectile moving through the material and exiting from the rear surface. Perforation is synonymous with complete penetration

**PETN**  A white crystalline explosive solid—Pentaerythritol Tetranitrate, with the chemical formula $C_5H_8N_4O_{12}$

**Phase**  A distinct state of matter in a system

**Plugging**  With reference to a projectile impacting and penetrating a target, if the material is susceptible to shear failure, a plug can be detached from the armour. This forms a secondary projectile that can result in catastrophic behind-armour effects. If plugging failure does occur, the total energy that the armour absorbs will be less. This is because the failure is localised and does not allow for gross plate plastic deformation. In armour materials, plugging is usually a result of adiabatic shear

**Polycrystalline**  With reference to a material, a structure that consist of multiple crystals (or 'grains') that are joined together, usually with random orientation

**Primary explosive**  A sensitive explosive that is generally used in initiating devices

**Proof stress**  An arbitrary yield stress calculated for materials that do not have an obvious yield point—such as aluminium. A line is drawn parallel to the linear elastic part of the stress-strain curve but is offset by some standard amount of strain (for example, 0.1 %). The intersection of the offset line and the stress-strain curve is the proof stress. Sometimes referred to as the *offset yield stress*

**Propellant**  An explosive that burns to propel a projectile or missile through the expansion of high-pressure gases

**Serpentin**  A lever that carries a smouldering wick (or match) on a matchlock weapon

**Quenching**  The process of hardening steel by rapidly cooling it from some elevated temperature. Water is the most commonly used medium for quenching although oils are sometimes used

**Radial fracture**  Cracking that resembles the pattern of spokes in a bicycle wheel

**Rolled homogeneous armour**  Rolled steel plate for armour applications that possesses relatively high hardness and good through-thickness properties. It usually contains of carbon, manganese, nickel, and molybdenum

**Sabot**  A light-weight full-calibre casing designed to carry a sub-calibre projectile up the gun tube. It is usually discarded after the projectile leaves the muzzle

**Scalar**  A quantity that has magnitude only (no direction)

**Secondary explosive**  An explosive material that is relatively insensitive compared to a primary explosive and usually requires an initiating device to cause detonation

**Secondary penetration** Penetration that occurs in a target after a long-rod penetrator has been completely eroded

**Self-forging fragment** See *explosively-formed projectile*

**Shock wave** A propagating discontinuity of pressure, temperature (or internal energy), and density that is spread over a very thin front

**Slipping driving band** A malleable band that is able to spin around the rim of a projectile or sabot to minimise the spin imparted by the rifling of a gun barrel. It is used for launching drag-stabilised projectiles in rifled gun barrels

**Spall** Fast tensile failure in a material

**Spall liner** A sheet of material that is applied on the inside of a vehicle hull to minimise the lethal effects of a perforating projectile or jet

**Strain** The deformation of a material

**Strain hardening** See *work hardening*

**Strain rate** The rate at which deformation occurs

**Stress** Force divided by the cross-sectional area over which the force is applied

**Stress corrosion cracking** A failure that occurs when a metal is attacked by a corrodant whilst being subjected to tensile stress. It is a particularly insidious failure because the magnitudes of stresses that are required to encourage failure are frequently lower than the yield strengths of the metal. In fact, residual stresses induced during machining, assembling or welding can exacerbate this type of failure. Aluminium alloys used in some armoured fighting vehicles were susceptible to this type of failure

**Sub calibre** Where the diameter of the projectile that strikes that target is smaller than the calibre of the gun that fired it. The projectile would have to be carried by some form of sabot

**Super calibre** Referring to a projectile where the bore diameter of the gun is smaller than the projectile launched from it

**Tempering** A process of heating a metal to relieve internal stresses and to produce the desired quantities of hardness and toughness

**TNT** A pale yellow solid, Trinitrotoluene, with the chemical formula $C_7H_5N_3O_6$

**Ultimate tensile strength** The maximum stress as measured from an engineering stress-strain curve

**Vector** A quantity that has both magnitude and direction

**Wire-wound gun** A gun that is built up using a tube that serves as the bore onto which a high tensile steel wire is wrapped and then jacketed. This was a very common technique for manufacturing Naval guns at the beginning of the 20th C

**Work hardening** Hardening of a material due to the application of work. This process is due to the presence of lattice dislocation pile-up and is sometimes referred to as *strain hardening*

**Yield strength** See *yield stress*

**Yield stress** The stress at which the material ceases to behave elastically. Also referred to as the *yield strength*

# Index

© Springer Nature Switzerland AG 2021
P. J. Hazell, *The Story of the Gun,* Springer Praxis Books,
https://doi.org/10.1007/978-3-030-73652-1

Printed in the United States
by Baker & Taylor Publisher Services